CAMBRIDGE TRACTS IN MATHEMATICS

73. *Degree theory*

N. G. LLOYD

Senior Lecturer in Pure Mathematics
The University College of Wales, Aberystwyth

Degree theory

CAMBRIDGE UNIVERSITY PRESS

CAMBRIDGE

LONDON · NEW YORK · MELBOURNE

Published by the Syndics of the Cambridge University Press
The Pitt Building, Trumpington Street, Cambridge CB2 1RP
Bentley House, 200 Euston Road, London NW1 2DB
32 East 57th Street, New York, NY 10022, USA
296 Beaconsfield Parade, Middle Park, Melbourne 3206, Australia

First published 1978

Printed in Great Britain
at the University Press, Cambridge

Library of Congress cataloguing in publication data

Lloyd, Noel Glynne, 1946–
Degree theory

(Cambridge tracts in mathematics; 73)
Bibliography: p. 165
Includes index
1. Degree, topological 2. Differential equations
numerical solutions 3. Functional analysis I. Title
II. Series
QA612.L57 515'.7 77–3205

ISBN 0 521 21614 1

Contents

Preface

Many problems in analysis and in the applications of analysis can be reduced to a study of the set of solutions of an equation of the form $f(x) = p$ in an appropriate space. Degree theory has developed as a means of examining this solution set and obtaining information on the existence of solutions, their number and their nature. The theory is widely used in the study of both ordinary and partial differential equations, and in that of more general functional equations. It is useful, for example, in bifurcation theory and in proving the existence of periodic solutions of differential equations. Several of these applications involve the use of convenient fixed point theorems; degree provides a natural technique for seeking such theorems, and, indeed, an array of them can be so obtained.

Suppose that D is an open subset of some topological space X, p is a point of X and f is a continuous function from D into X. The aim of degree theory is to define an integer $d(f, D, p)$, the degree of f at p relative to D, with the properties that (1) $d(f, D, p)$ is an estimate of the number of solutions of $f(x) = p$ in D, (2) d is continuous in both f and p, and (3) d is additive in the domain D in the sense that, if D is the union of disjoint open sets D_1 and D_2, then

$$d(f, D, p) = d(f, D_1, p) + d(f, D_2, p).$$

It is too much to hope for a successful definition in a completely general setting. The drift of this monograph is first to define $d(f, D, p)$ in the simplest case – when $X = \mathbf{R}^n$ and f is continuously differentiable – and then gradually to extend the classes of spaces and functions covered. As the spaces are generalised, more restrictions have usually to be placed on the functions. The tension between the generality of the space X and the restrictions on the function f provides some of the interest in the later developments. To distinguish this usage of the word degree from the many others in mathematics, it is often called 'topological degree'.

This book is designed as a guide to degree theory for analysts and applied mathematicians. The interest of the subject ultimately depends in large part on its applications – it is a useful component of the armoury of researchers into many of the applications of analysis. This I have sought to emphasise throughout; in the final (and longest) chapter some of the uses to which the theory has been put are illustrated.

The notion of the degree of a mapping was introduced by algebraic topologists. The approach here, however, is analytic. This better suits the inclinations of analysts, gives them more insight, and is more attuned to the applications in which applied mathematicians are likely to be interested. The treatment is elementary in that no previous knowledge of a specialised nature is assumed. The proofs of some of the early results involve some quite complicated analysis; those readers who find these proofs somewhat heavy going should by-pass them. A connection with differential topology is sketched in Section 2.4; it is not the purpose of the book to enter into its details.

A short course on degree theory could reasonably consist of the first four chapters (excluding Sections 2.4 and 3.3), followed by Chapter 6, and Chapter 9 for the applications. Well over half of Chapter 9 can be read immediately after Section 3.2.

The first two chapters are concerned with establishing the definition and properties of degree in finite dimensional spaces. This is sometimes called the Brouwer degree – the idea was first developed by Brouwer (in 1912). The analytic viewpoint was given by Nagumo (1951*a*). The treatment given here is that of Schwartz (1969). The approach from algebraic topology is covered in the book of Cronin (1964), who also gives a number of applications. In the third chapter, a few of the topological results which can be proved using the Brouwer degree are given; included are the Brouwer fixed point theorem and the higher dimensional analogue of the Jordan curve theorem.

In Chapter 4 the notion of degree is considered when the underlying space is an infinite dimensional normed space. The degree of compact perturbations of the identity in Banach spaces was defined by Leray and Schauder in a celebrated paper in 1934. This form of degree is, naturally, called the Leray–Schauder degree. When seeking to define the degree of successively larger classes of mappings, it is very useful to have at hand an axiomatic scheme; then only a small number of properties have to be checked. Such a scheme was given by Amann and Weiss (1973). This is the subject matter of Chapter 5. Compact mappings are sometimes too restrictive for applications. A useful generalisation is the idea of a strict set contraction. Mappings of this kind are investigated in Chapter 6, and the degree of perturbations of the identity by strict set contractions is defined. The axiomatic scheme of Chapter 5 proves to be particularly useful.

In their investigation of various classes of functional equations, Browder and his associates considered generalisations of compact mappings wider than strict set contractions. The degree of some of these is defined in Chapter 7 (Sections 1 and 2). A selection has to be made of which generalised degrees to include. One that is not discussed is the 'coincidence degree' of Mawhin (1972); this is a multivalued degree, and has been used in a number of problems in differential and functional-differential equations. The last section of Chapter 7 deals with mappings which are multivalued. In Chapter 7 as a whole comparatively few details are supplied. Other generalisations of the Leray–Schauder degree have been developed, using Smale's version of Sard's theorem. This is an instructive approach, especially when it is desired to calculate degrees; it is sketched in Chapter 8.

The final chapter is designed as a survey of some of the applications of degree theory, particularly in the theory of differential equations. The whole field of fixed point theorems is dealt with at various places in the previous chapters. Inevitably, in a short space, it is possible only to select a few areas and to try to give the reader some feeling of their flavour. The particular classes of equations discussed and the particular results proved for them have been selected with their illustrative merits in mind.

The list of references is not intended as a complete bibliography of the subject – particularly in the case of applications; it is confined to those writings to which reference is made in the text. It should be noted that, for fixed point theorems, this book is complementary to that of Smart (1974).

When dealing with infinite dimensional spaces, I have restricted attention to normed spaces. Most of the material goes over virtually unchanged to the case of more general topological vector spaces (locally convex, say). The interested reader is likely to have little difficulty in doing the appropriate adaptations.

The book has grown out of a course of lectures which I gave in the University of Cambridge in 1972 and 1973. My own understanding of the subject was much assisted by the lectures given by Professor Edward Fraenkel in Cambridge in 1973.

I am most grateful to Dr T. W. Körner, Dr J. B. McLeod and Professor Sir Peter Swinnerton-Dyer for reading the typescript and making many helpful suggestions. I am also grateful to Mrs Noreen Davies for her patient typing of the manuscript and to the staff of Cambridge University Press for their care during publication.

N. G. L.

Aberystwyth
October 1976

1
Degree theory in finite dimensional spaces

This first chapter is devoted to defining the degree of continuous functions defined on subsets of $\mathbf{R}^n$; there is then no difficulty in extending the definition to apply in any finite dimensional normed space. We have to proceed to the definition in a number of stages. Our treatment closely follows that of Schwartz (1969), and is related to that of Nagumo (1951*a*).

In this and the subsequent two chapters we use the following notation:

(i) $x = (x_1, \ldots, x_n) \in \mathbf{R}^n$; $|x| = \max\{|x_i|; i = 1, \ldots, n\}$.

(ii) Unless the contrary is explicitly stated, D is a bounded, open subset of $\mathbf{R}^n$ and p is a point of $\mathbf{R}^n$. Given a set S, its closure is written $\bar{S}$, its interior int S, its boundary ∂S and its complement $\mathscr{C}S$.

(iii) $C(\bar{D})$ is the linear space of continuous functions from $\bar{D}$ into $\mathbf{R}^n$ with the norm

$$\|f\| = \sup_{x \in D} |f(x)|.$$

(iv) $f'(x)$ is the derivative of the function f at the point x; the symbol $,_j$ denotes the operator $\partial/\partial x_j$; $J_f(x)$ is the Jacobian determinant of f at x:

$$J_f(x) = \det f'(x) = \det\,(f_{i,j}(x)).$$

(v) $C^1(\bar{D})$ is the space defined as follows: $f \in C^1(\bar{D})$ if $f \in C(\bar{D})$ and there is an extension $\tilde{f}$ of f defined on an open set $\mathscr{D}(f)$ containing $\bar{D}$ such that f has continuous first order partial derivatives in $\mathscr{D}(f)$; the norm on $C^1(\bar{D})$ is

$$\|f\|_1 = \sup_{\substack{x \in D \\ 1 \leqslant i \leqslant n}} |f_i(x)| + \sup_{\substack{x \in D \\ 1 \leqslant i, j \leqslant n}} |f_{i,j}(x)|.$$

(vi) $C^2(\bar{D})$ is the subspace of $C^1(\bar{D})$ consisting of those functions for which the corresponding extensions $\tilde{f}$ have continuous partial derivatives of the second order in $\mathscr{D}(f)$.

(vii) $C_0^r(\bar{D})(r=1,2)$ is the subspace of $C^r(\bar{D})$ consisting of those functions whose supports are contained in D. The *support* of f, supp f, is $\overline{\{x;f(x)\neq 0\}}$.

(viii) A point x for which $f(x)=p$ is called a *p-point* of f. For $f\in C(\bar{D})$, $f^{-1}(p)$ is then the collection of p-points of f in $\bar{D}$.

Later in the book we discuss infinite dimensional normed spaces $(X, \|.\|)$. We retain the notations (ii), (iii) and (viii) with X replacing $\mathbf{R}^n$ and $\|.\|$ replacing $|.|$. Whatever the norm under consideration, ρ denotes the induced distance (induced by $|.|$ in $\mathbf{R}^n$, by $\|.\|$ in infinite dimensional spaces). $B(y,r)$ is the open ball centre y, radius r:

$$B(y,r)=\{x;\rho(x,y)<r\};$$

$\bar{B}(y,r)$ is the closed ball $\{x;\rho(x,y)\leqslant r\}$.

1.1 Definition of degree for C^1 functions

We start with some definitions.

Definition 1.1.1 Let $\phi\in C^1(\bar{D})$. We say that x is a *critical point* of ϕ if $J_\phi(x)=0$; then $\phi(x)$ is a *critical value* of ϕ. The set of critical points of ϕ in $\bar{D}$ is denoted by $Z_\phi(\bar{D})$, or simply Z_ϕ; the set of critical values, $\phi(Z_\phi)$, is called the *crease* of ϕ.

The word 'crease' is used to suggest the idea that at a critical point y, ϕ is not well approximated by a linear function – that is, ϕ does not locally 'look' linear; ϕ need not be one to one in a neighbourhood of y. The inverse image of a crease point may be an infinite set; conversely, we have the following.

Theorem 1.1.2 *If $\phi\in C^1(\bar{D})$ and $p\notin$ crease ϕ, then $\phi^{-1}(p)$ is finite.*

Proof Since $\bar{D}$ is compact, the result follows if we show that the p-points of ϕ in $\bar{D}$ are isolated. If this were not so, we should have a sequence (x_n) in $\bar{D}$ convergent to x_0, say, with $\phi(x_n)=p$ for all n. Then $x_0\in\bar{D}$, $\phi(x_0)=p$, and

$$0 = \phi(x_n) - \phi(x_0) = \phi'(x_0)(x_n - x_0) + o(|x_n - x_0|) \quad (n \to \infty). \tag{1.1.1}$$

Since $\phi'(x_0)$ is non-singular, there is $r > 0$ such that

$$|\phi'(x_0)u| \geqslant r|u| \quad (u \in \mathbf{R}^n). \tag{1.1.2}$$

But, for sufficiently large n, (1.1.1) implies that

$$|\phi'(x_0)(x_n - x_0)| < \tfrac{1}{2}r|x_n - x_0|,$$

contradicting (1.1.2). The result follows.

We can now define the degree of ϕ at p when ϕ is a C^1 function and p is not a critical value of ϕ. It is the algebraic number of p-points in D; it counts $+1$ or -1 at a p-point x according as ϕ is orientation preserving or orientation reversing near x.

Definition 1.1.3 Suppose that $\phi \in C^1(\bar{D})$, $p \notin \phi(\partial D)$ and $p \notin$ crease ϕ. Define the *degree of ϕ at p relative to D* to be $d(\phi, D, p)$, where

$$d(\phi, D, p) = \sum_{x \in \phi^{-1}(p)} \operatorname{sign} J_\phi(x).$$

Remarks (1) Since $p \notin$ crease ϕ, Theorem 1.1.2 implies that the summation in the definition is finite.

(2) The condition $p \notin \phi(\partial D)$ is essential; it cannot be removed.

(3) We shall investigate the dependence of $d(\phi, D, p)$ on each of ϕ, D and p; for example, we shall see that d is continuous in ϕ and p when D is fixed.

(4) There is a means of defining degree relative to an unbounded set D; an indication of this was given by Nagumo (1951*a*).

(5) 'Degree' is often termed 'topological degree'.

The following is clear from Definition 1.1.3.

Theorem 1.1.4 *Let I denote the identity mapping. If $p \in D$, then $d(I, D, p) = 1$; if $p \notin \bar{D}$, then $d(I, D, p) = 0$.*

The task ahead is to remove the restrictions $\phi \in C^1$, $p \notin$ crease ϕ imposed in Definition 1.1.3. This is done by a process of approximation. We start by proving that, if $\phi \in C^1(\bar{D})$, $p \notin \text{crease}\,\phi$, $p \notin \phi(\partial D)$, and ψ is sufficiently near ϕ (in the C^1 topology), then $d(\psi, D, p)$ is defined and is equal to $d(\phi, D, p)$. The proof which we give is due to L. E. Fraenkel (1973); it is didactically preferable to a somewhat simpler line of proof which uses Theorem 1.3.1. It is shown, using the contraction mapping principle, that the set $\phi^{-1}(p)$ is structurally stable in the sense that a small change in ϕ merely produces a small shift of the p-points, and does not alter their number.

Theorem 1.1.5 *Suppose that $\phi \in C^1(\bar{D})$ and $p \notin \phi(Z_\phi) \cup \phi(\partial D)$. There is $\delta > 0$, depending on p and ϕ, such that, if $\| \psi - \phi \|_1 < \delta$, then $p \notin \psi(Z_\psi) \cup \psi(\partial D)$ and $d(\psi, D, p) = d(\phi, D, p)$.*

Proof If $\phi^{-1}(p) = \emptyset$, take $\delta = \frac{1}{2}\rho(p, \phi(\bar{D}))$ (where ρ, as indicated at the beginning of the chapter, is the distance induced by $|\cdot|$ on $\mathbf{R}^n$). Then, if $\| \psi - \phi \|_1 < \delta$,

$$|\psi(x) - p| > \delta \quad (x \in \bar{D}).$$

Hence ψ has no p-points in $\bar{D}$, and, by Definition 1.1.3,

$$d(\phi, D, p) = d(\psi, D, p) = 0.$$

Suppose now that $\phi^{-1}(p) = \{a_1, \ldots, a_k\}$ – it is known that $\phi^{-1}(p)$ is finite, for p is not a critical value of ϕ. We take disjoint balls $B_i = B(a_i, r)$ and make a sequence of choices of r and δ so that eventually every $\psi \in C^1(\bar{D})$ satisfying $\| \psi - \phi \|_1 < \delta$ has exactly one p-point in each B_i and none elsewhere.

First choose $r = r_0$ so that the $\bar{B}_i$ are disjoint balls not meeting ∂D and Z_ϕ. Let $B(r) = B(a_1, r) \cup \ldots \cup B(a_k, r)$ and

$$c = \min_{i=1 \ldots k} |J_\phi(a_i)|.$$

So choose $r_1 \leqslant r_0$ that $|J_\phi(x)| \geqslant \frac{2}{3}c$ if $x \in B(r_1)$; then choose δ_1 so that $\| \psi - \phi \|_1 \leqslant \delta_1$ implies

$$|J_\phi(x) - J_\psi(x)| \leqslant \tfrac{1}{3}c \quad (x \in \bar{D}).$$

It follows that $|J_\psi(x)| \geqslant \frac{1}{3}c$ if $x \in B(r_1)$; in particular, no point within r_1 of any a_i is a critical point of ψ.

We proceed to use the contraction mapping theorem to prove that ψ has exactly one p-point in each B_i. Consider a fixed index i, and write

$$a_i = a, \quad x - a = z, \quad \phi(a) - \psi(a) = h.$$

We therefore seek to solve

$$\psi(x) - \psi(a) = h$$

in $B \equiv B_i$. Since ψ is a C^1 function, this can be written

$$\psi'(a)z + T(z) = h, \tag{1.1.3}$$

where $T(z) = \psi(z+a) - \psi(a) - \psi'(a)z$. Now

$$\frac{d}{d\theta}\psi_i(a + \theta z + (1-\theta)y) = \sum_{j=1}^{n}(z_j - y_j)(\psi_{i,j}(a + \theta z + (1-\theta)y)),$$

so that

$$(T(z) - T(y))_i = \sum_{j=1}^{n}(z_j - y_j)\int_0^1 \{\psi_{i,j}(a + \theta z + (1-\theta)y) - \psi_{i,j}(a)\}\,d\theta. \tag{1.1.4}$$

It was arranged that $|J_\psi| \geqslant \frac{1}{3}c$ in $B(r_1)$; therefore $\psi'(a)$ is an invertible linear operator – let its inverse be V, and let $|V| = \sup\{|Vx|; |x| = 1\}$ be its norm. Then $|Vh| \leqslant |V|\delta_1$, and (1.1.3) becomes

$$V(h - T(z)) = z. \tag{1.1.5}$$

We shall write $W(z)$ for $V(h - T(z))$; we seek the fixed points of W in $B(0, r)$.

If

$$\|\phi - \psi\|_1 \leqslant \delta \leqslant \delta_1 \quad \text{and} \quad |z|, |y| \leqslant r \leqslant r_1,$$

then manipulation of (1.1.4) using the triangle inequality gives

$$\begin{aligned} |T(z) - T(y)| &= |z - y|[O(\delta) + O(r)] \\ &\leqslant |z - y| K(\delta + r), \end{aligned}$$

say, where K is a positive constant independent of δ and r, but depending, of course, on ϕ and a. Also

$$|W(z)-W(y)| \leqslant |z-y|K(\delta+r)|V|, \tag{1.1.6}$$

and, putting $y=0$,

$$|W(z)| \leqslant (Kr(\delta+r)+\delta)|V|. \tag{1.1.7}$$

Now choose $r \leqslant r_1$ so that $Kr|V|<\frac{1}{3}$; then choose $\delta \leqslant \delta_1$ so that $\delta|V|<\frac{1}{3}r$ and $K\delta|V|<\frac{1}{3}$. From (1.1.6) and (1.1.7), we see that W maps $B(0,r)$ into itself, and is a contraction there. Consequently ψ has exactly one p-point in B. The same argument applied to each of the balls $B_1,\ldots,B_k$ yields finally the existence of $\delta>0$ and $r>0$ such that ψ has exactly one p-point in each of $B(a_i,r)$ if $\|\psi-\phi\|_1<\delta$. We note that J_ϕ and J_ψ have the same sign in each $B(a_i,r)$.

With r as chosen above, there is $\varepsilon>0$ such that $|\phi(x)-p| \geqslant \varepsilon$ for $x \in F \equiv \bar{D}\backslash B(r)$. By decreasing δ further if necessary, we can ensure that $|\phi(x)-\psi(x)| \leqslant \frac{1}{2}\varepsilon$ if $x \in \bar{D}$. Then, if $x \in F$, $|\psi(x)-p| \geqslant \frac{1}{2}\varepsilon$, whence $\psi^{-1}(p) \subset B(r)$. Since no point of $B(r)$ is a critical point of ψ, we also have $p \notin$ crease ψ. Finally

$$\begin{aligned} d(\psi,D,p) &= \sum_{x\in\psi^{-1}(p)} \operatorname{sign} J_\psi(x) \\ &= \sum_{i=1}^{k} \operatorname{sign} J_\phi(a_i) \\ &= d(\phi,D,p), \end{aligned}$$

which is the required conclusion.

The last result is at the root of much that follows. It is a restricted form of a theorem we shall encounter later.

Before starting the process of extending the scope of our definition, we briefly describe a few of the main features of degree which will be proved later.

(1) Suppose that D is the disjoint union of open sets $D_1,\ldots,D_m$. When $\phi \in C^1(\bar{D})$ and p is not a crease point of ϕ, it is obvious from Definition 1.1.3 that

$$d(\phi,D,p)=\sum_{i=1}^{m} d(\phi,D_i,p).$$

It is not surprising that the same formula holds if $\phi \in C(\bar{D})$ and

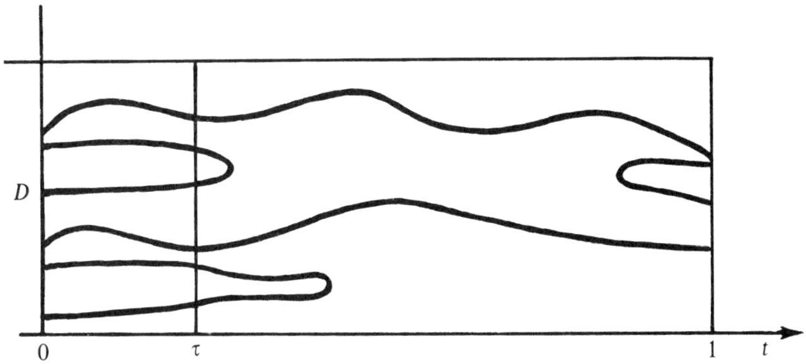

Figure 1.1. Homotopy invariance.

$p \in$ crease ϕ; we shall also find that the formula holds when D is the union of infinitely many open sets.

(2) If ϕ, ψ are C^1 functions and $p \notin$ crease ϕ, we have seen that the sets $\phi^{-1}(p)$ and $\psi^{-1}(p)$ are close to each other when $\| \phi - \psi \|_1$ is small enough. This property will be generalised: $d(\phi, D, p) = d(\psi, D, p)$ if $\| \phi - \psi \|$ is small enough. This is intuitively clearer if it is rephrased: '$d(\phi, D, p)$ is continuous in ϕ'. It will also be seen that d is continuous in p – another fact which is fairly obvious when ϕ and p are restricted as in Definition 1.1.3.

(3) One of the characteristic properties of degree is 'homotopy invariance'. Suppose that ϕ undergoes a continuous deformation (to ψ, say) in the course of which no p-points appear on the boundary ∂D. Let the deformation, or homotopy, be given by the continuous mapping H: $[0, 1] \times \bar{D} \to \mathbf{R}^n$; let $h_t(x) = H(t, x)$, so that $h_0 = \phi$ and $h_1 = \psi$. Without entering into the details, the inverse image $H^{-1}(p)$ is, for a dense set of p, the union of curves in $[0, 1] \times \bar{D}$; these curves do not meet $[0, 1] \times \partial D$. Diagrammatically, we have something like Figure 1.1; $h_\tau^{-1}(p)$ is represented by the points of the curves on the vertical line $t = \tau$. The diagram indicates that as t varies, p-points of h_t appear and disappear in pairs; the Jacobian of h_t has opposite signs at the two members of a pair. The number of p-points changes when p becomes a critical value of h_t. In the light of these remarks, it is not surprising that, if $p \notin h_t(\partial D)$ for $0 \leqslant t \leqslant 1$, then $d(\phi, D, p) = d(\psi, D, p)$. This, the property of invariance under homotopy, we shall prove later.

(4) Under the conditions of Definition 1.1.3, if $d(\phi, D, p) \neq 0$, then D must contain a p-point of ϕ. This property is generally true, and is very useful. Indeed, degree theory would lose much of its usefulness if the conclusion were ever false.

The reader may find it useful at this juncture to look at the simple illustrations given in Section 1.4, verifying in each case the four properties mentioned above.

1.2 Towards a definition for continuous functions

The sequence of definitions of degree which we follow may be summarised in this way:

(i) $\phi \in C^1, p \notin \phi(Z)$, (ii) $\phi \in C^2$, any p, (iii) $\phi \in C^1$, any p, (iv) $\phi \in C$.

The step (i) was, of course, our starting point; the restriction on p to be non-critical is removed at the expense of having to suppose ϕ to be twice continuously differentiable. Steps (iii) and (iv) involve, respectively, approximating a C^1 function with C^2 functions and approximating to a continuous function with C^1 functions.

To complete this programme, a number of somewhat technical lemmas are required, together with a simple version of Sard's theorem (Sard, 1942). This section is a recital of these results; the reader may well wish to avoid their proofs, and proceed immediately to Section 1.3. Sard's theorem states that the crease of a C^1 function is 'small'.

Theorem 1.2.1 *Suppose that $\phi \in C^1(\bar{D})$; the crease $\phi(Z_\phi)$ has zero measure in $\mathbf{R}^n$.*

Proof For $x \in D$, let $\eta_x = \frac{1}{2}\rho(x, \mathscr{C}D)$ and $R(x)$ the open cube of side η_x centred at x. Then $R(x) \subset D$ and $D = \bigcup_{x \in D} R(x)$; hence, by the Lindelöf property of $\mathbf{R}^n$, D is the union of countably many of the $R(x)$:

$$D = \bigcup_{i=1}^{\infty} R(x_i).$$

Thus it is necessary only to show that the result holds when $D = R$, a cube. Suppose, then, that R is a cube of side l. If $\phi = (\phi_1, \dots, \phi_n)$, we have, for $x, y \in \mathbf{R}^n$ and $1 \leqslant i \leqslant n$,

$$\phi_i(y) - \phi_i(x) = \sum_{j=1}^{n} \phi_{i,j}(z^{(i)})(y_j - x_j), \tag{1.2.1}$$

where the $z^{(i)}$ are points on the line segment joining x and y. So

$$|\phi(y) - \phi(x)| \leqslant M|y - x| \tag{1.2.2}$$

for some constant M. Define

$$T_i^x(y) = \phi_i(x) + \sum_{j=1}^{n} \phi_{i,j}(x)(y_j - x_j) \quad (i = 1, \dots, n). \tag{1.2.3}$$

T^x is the affine map approximating ϕ at x. Then, by (1.2.1) and (1.2.3),

$$\phi_i(y) - T_i^x(y) = \sum_{j=1}^{n} (\phi_{i,j}(z^{(i)}) - \phi_{i,j}(x))(y_j - x_j) \quad (i = 1, \dots, n).$$

Since $\phi \in C^1(\bar{R})$ this implies that, given $\varepsilon > 0$, there is $\delta > 0$ such that

$$|\phi(y) - T^x(y)| \leqslant \varepsilon|x - y| \quad (|x - y| < \delta). \tag{1.2.4}$$

Suppose that $x \in Z_\phi$; then $\det(T^x) = 0$, whence T^x maps $\mathbf{R}^n$ into an affine subspace P^x of $\mathbf{R}^n$ of dimension at most $n - 1$. If $|x - y| < \eta < \delta$, then (1.2.2) tells us that $|\phi(x) - \phi(y)| \leqslant M\eta$, and (1.2.4) implies that $\rho(\phi(y), P^x) \leqslant \varepsilon\eta$. This means that if $|x - y| < \eta$, $\phi(y)$ is contained in a cuboid centred at $\phi(x)$ with sides $2M\eta$ in P^x and $2\varepsilon\eta$ perpendicular to P^x. The volume of this cuboid is $2^n M^{n-1} \varepsilon \eta^n$.

Now divide R into s^n cubes $R_m (m = 1, \dots, s^n)$ each of side l/s, s being so chosen that $l < \delta s$. (Recall that l is the length of the side of R.) If $x \in R_t \cap Z_\phi$, certainly $R_t \subset \{y;\ |y - x| \leqslant l/s\}$; therefore $\phi(R_t)$ has measure less than $2^n M^{n-1} \varepsilon (l/s)^n$. Consequently

$$\text{measure } [\phi(Z_\phi)] \leqslant (2l)^n M^{n-1} \varepsilon.$$

But ε was chosen arbitrarily, and l and M are independent of ε, whence

$$\text{measure}\,[\phi(Z_\phi)] = 0,$$

as required.

Remark Without going into details, we note that there is a much more general form of Sard's theorem, which states that, if M, N are C^k manifolds of dimensions m, n, respectively, and $f: M \to N$ is C^k, then the critical values of f form a set of measure zero if $k - 1 \geqslant \max(m - n, 0)$. For a proof, see Milnor (1965) or Sternberg (1964). Smale (1965) has proved a version applicable to manifolds modelled on Banach spaces.

We now come to a sequence of lemmas which will be required in the next section. These tell us that certain functions are divergences. We write $\sum_{i=1}^{n} u_{i,i} = \operatorname{div} u$. If X is a subset of $\mathbf{R}^n$, it is convenient to write $C^1(X, \mathbf{R}^m)$ for the vector space of continuously differentiable functions from X into $\mathbf{R}^m$, and $C_0^1(X, \mathbf{R}^m)$ for the subset consisting of those functions whose supports are compact. Generally $X = \mathbf{R}^n$ or $X = \bar{D}$.

Lemma 1.2.2 *Let* $\phi \in C^2(\bar{D})$ *and* $v \in C_0^1(\mathbf{R}^n, \mathbf{R}^n)$ *with* supp v *disjoint from* $\phi(\partial D)$. *Then there is a function* $u \in C_0^1(\bar{D})$ *such that*

$$(\operatorname{div} u)(x) = J_\phi(x) \cdot (\operatorname{div} v)(\phi(x)).$$

Proof Let $A_{ji}(x)$ be the cofactor of $\phi_{j,i}$ in J_ϕ; define

$$u_i(x) = \sum_{j=1}^{n} v_j(\phi(x)) A_{ji}(x).$$

Clearly $u \in C_0^1(\bar{D})$, and we have

$$(\operatorname{div} u)(x) = \sum_{i,j,k=1}^{n} v_{j,k}(\phi(x)) \phi_{k,i}(x) A_{ji}(x) + \sum_{i,j=1}^{n} v_j(\phi(x)) A_{ji,i}(x).$$

Now

$$\sum_{k=1}^{n} \phi_{k,i} A_{ji} = \delta_{kj} J_\phi \tag{1.2.5}$$

and, for $i = 1, \ldots, n$,

$$\sum_{j=1}^{n} A_{ij},_j = 0. \tag{1.2.6}$$

It follows that $\quad (\text{div } u)(x) = \sum_{j=1}^{n} v_j,_j(\phi(x)) J_\phi(x),$

as required. The identity (1.2.5) is immediate from the definition of A_{ij}; to prove (1.2.6), fix i and let g be the column vector

$$\begin{bmatrix} \phi_1 \\ \vdots \\ \hat{\phi}_i \\ \vdots \\ \phi_n \end{bmatrix},$$

where $\hat{}$ denotes the omission of a component. Also let

$$G_j = \det[g,_1, \ldots, \hat{g},_j, \ldots, g,_n]$$

and $$\Gamma_{pj} = \begin{cases} \det[g,_j,_p, g,_1, \ldots, \hat{g},_p, \ldots, \hat{g},_j, \ldots, g,_n] & \text{if } p < j \\ \det[g,_j,_p, g,_1, \ldots, \hat{g},_j, \ldots, \hat{g},_p, \ldots, g,_n] & \text{if } p > j. \end{cases}$$

Then $$G_j,_j = \sum_{p<j} (-1)^{p+1} \Gamma_{pj} + \sum_{p>j} (-1)^p \Gamma_{pj},$$

whence $$\sum_{j=1}^{n} (-1)^j G_j,_j = 0, \quad \text{and} \quad \sum_{j=1}^{n} A_{ij},_j = 0.$$

Lemma 1.2.3 *Let $f \in C_0^1(\bar{D}, \mathbf{R}^1)$ and $K = \text{supp } f$. Suppose that for some $\bar{x} \in \mathbf{R}^n$,*

$$A \equiv \{k + \theta\bar{x}; k \in K, 0 \leqslant \theta \leqslant 1\} \subset D.$$

Then there is $v \in C_0^1(\bar{D})$ such that

$$(\text{div } v)(x) = f(x) - f(x - \bar{x}).$$

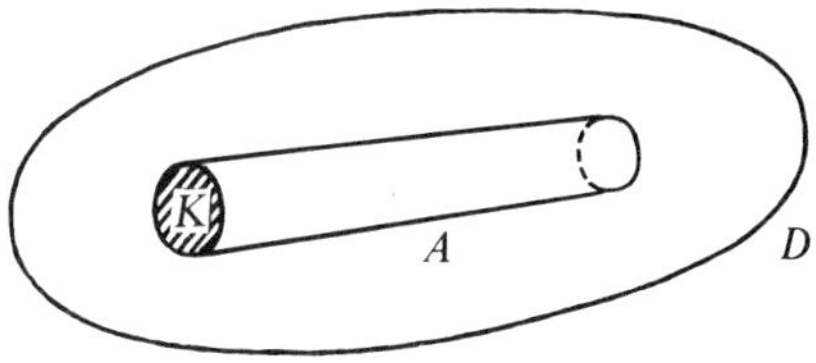

Figure 1.2. The set A of Lemma 1.2.3

Proof The set A is illustrated in Figure 1.2. Define $F:\bar{D}\to \mathbf{R}^1$ and $v:\bar{D}\to \mathbf{R}^n$ by

$$F(x)=\int_0^1 f(x-\theta\bar{x})\,d\theta,\quad v(x)=(F(x))\bar{x}.$$

Then $v\in C^1(\mathbf{R}^n,\mathbf{R}^n)$. Moreover, since $f(x-\theta\bar{x})=0$ if $x-\theta\bar{x}\notin K$, $F(x)=0$ if $x\notin \bigcup_{0\leqslant\theta\leqslant 1}(K+\theta\bar{x})$; thus supp $F\subset A$, and $v\in C_0^1(\bar{D})$.

Also
$$\begin{aligned}(\operatorname{div} v)(x) &= \sum_{i=1}^n \bar{x}_i F_{,i}(x)\\ &= \int_0^1 \sum_{i=1}^n \bar{x}_i f_{,i}(x-\theta\bar{x})\,d\theta\\ &= -\int_0^1 \frac{d}{d\theta} f(x-\theta\bar{x})\,d\theta\\ &= f(x)-f(x-\bar{x}).\end{aligned}$$

Lemma 1.2.4 *Let $f\in C_0^1(\mathbf{R}^n,\mathbf{R})$ and $K=\operatorname{supp} f$. Let $\gamma(s)$ be a path in $\mathbf{R}^n$ such that the tube $A\equiv\{k+\gamma(s);\,k\in K,\,0\leqslant s\leqslant 1\}$ is contained in D. Then there is $v\in C_0^1(\bar{D})$ such that*

$$(\operatorname{div} v)(x)=f(x-\gamma(0))-f(x-\gamma(1)).$$

(Note that Lemma 1.2.3 is concerned with a particular path.)

Proof The relation between the various sets occurring is illustrated in Figure 1.3. For $s,t\in[0,1]$, we say that $s\sim t$ if

$$f(x-\gamma(s))-f(x-\gamma(t))$$

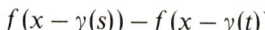

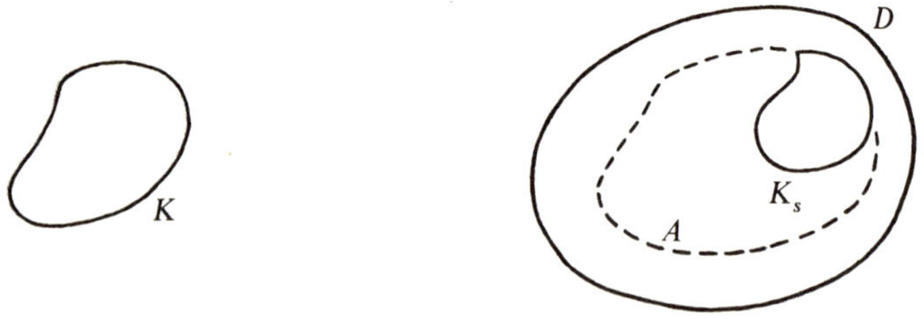

Figure 1.3. The set A of Lemma 1.2.4

is the divergence of a member of $C_0^1(\bar{D})$. It is easily seen that $\sim$ is an equivalence relation. We shall show that every equivalence class is open, so that, by the connectedness of $[0, 1]$, there is only one such class. Take a fixed s. Let $K_s = \{k + \gamma(s); k \in K\}$, $f_s(x) = f(x - \gamma(s))$ and $h_t = \gamma(t) - \gamma(s)$. Then $\operatorname{supp} f_s = K_s$. Since K_s is a compact subset of D, $\eta = \rho(K_s, \partial D) > 0$. So there is $\varepsilon > 0$ such that $|t - s| < \varepsilon$ implies $|h_t| < \frac{1}{2}\eta$, whence

$$K_s' \equiv \{k + \theta h_t\,; k \in K_s, 0 \leqslant \theta \leqslant 1\} \subset D.$$

By Lemma 1.2.3, there is $v \in C_0^1(\bar{D})$ such that if $|t - s| < \varepsilon$,

$$(\operatorname{div} v)(x) = f_s(x) - f_s(x - h_t) = f(x - \gamma(s)) - f(x - \gamma(t)).$$

Hence the equivalence classes are open, as required.

We are now in a position to progress with the definition of degree, and that we do in the next section.

1.3 Admitting critical values

So far we have insisted that p is not a critical value of ϕ. We now remove this restriction; this we accomplish by first expressing the degree of ϕ as an integral involving an averaging kernel. This integral could be taken as the definition of degree, an approach used by Heinz (1959).

Theorem 1.3.1 *Suppose that $\phi \in C^1(\bar{D})$, $p \notin \phi(\partial D)$ and $p \notin$ crease ϕ. Let $f_\varepsilon : \mathbf{R}^n \to \mathbf{R}$ be a continuous function such that*

$$K_\varepsilon \equiv \operatorname{supp} f_\varepsilon \subset B(0, \varepsilon), \qquad \int_{\mathbf{R}^n} f_\varepsilon(x)\, dx = 1.$$

There is ε_0, depending on p and ϕ, such that, if $0 < \varepsilon < \varepsilon_0$, then

$$d(\phi, D, p) = \int_D f_\varepsilon(\phi(x) - p)\, J_\phi(x)\, dx. \tag{1.3.1}$$

Proof First, functions f_ε with the above properties certainly exist. Since $p \notin$ crease ϕ, $\phi^{-1}(p)$ is finite: let $\phi^{-1}(p) = \{a_1, \ldots, a_k\}$. For sufficiently small ε, there are disjoint neighbourhoods $A_i(\varepsilon)$

of a_i in D such that, for each i, $\phi(A_i(\varepsilon)) = B(p, \varepsilon)$ and $\phi|_{A_i}$ is one to one. So choose ε that, in addition, the $A_i(\varepsilon)$ are disjoint from ∂D and in them $J_\phi \neq 0$. Then

$$\operatorname{supp} f_\varepsilon(\phi(.) - p) \subset \bigcup A_i.$$

Thus

$$\int_D f_\varepsilon(\phi(x) - p) J_\phi(x)\, dx = \sum_{i=1}^{k} \int_{A_i} f_\varepsilon(\phi(x) - p) J_\phi(x)\, dx.$$

But J_ϕ is of one sign in each $A_i(\varepsilon)$; hence

$$\int_{A_i} f_\varepsilon(\phi(x) - p) J_\phi(x)\, dx = \operatorname{sign} J_\phi(a_i) \int_{K_\varepsilon} f_\varepsilon(y)\, dy.$$

It follows that

$$\begin{aligned} \int_D f_\varepsilon(\phi(x) - p) J_\phi(x)\, dx &= \sum_{i=1}^{k} \operatorname{sign} J_\phi(a_i) \\ &= d(\phi, D, p). \end{aligned}$$

The next stage is to examine the effect on $d(\phi, D, p)$ of changes in p. It is clear from the proof of Theorem 1.1.5 that, if p_1 and p_2 are in the same component of $\mathbf{R}^n \backslash (\phi(\partial D) \cup \phi(Z_\phi))$, then $d(\phi, D, p_1) = d(\phi, D, p_2)$. The crease of ϕ acts as a barrier across which p cannot yet move without possibly changing the value of $d(\phi, D, p)$. The difficulty is that as p approaches the crease, the ε_0 of Theorem 1.3.1 may shrink to zero. We resolve the problem by using the lemmas of Section 1.2.

Theorem 1.3.2 *Let $\phi \in C^1(\bar{D})$. Suppose that p_1 and p_2 are in the same component of $\mathbf{R}^n \backslash \phi(\partial D)$, and that neither is in the crease of ϕ. Then*

$$d(\phi, D, p_1) = d(\phi, D, p_2).$$

Proof We shall prove the result first for C^2 functions.

(i) Suppose that $\phi \in C^2(\bar{D})$. $\mathbf{R}^n \backslash \phi(\partial D)$ is an open subset of $\mathbf{R}^n$; its (connected) components are therefore path-connected (see, for example, Newman (1939)). Let Ω be the component of $\mathscr{C}(\phi(\partial D))$

containing p_1 and p_2; there is a path $\gamma(s)$ in Ω with $\gamma(0)=p_1$ and $\gamma(1)=p_2$. Because $\{\gamma(s); 0 \leqslant s \leqslant 1\}$ is compact, there is $\varepsilon_1 > 0$ such that the ε_1-neighbourhood of γ is contained in Ω. Let $\varepsilon < \varepsilon_1$ and f_ε be so chosen that both $d(\phi, D, p_1)$ and $d(\phi, D, p_2)$ are given by the formula (1.3.1). Let

$$K_{s,\varepsilon} = \{z + \gamma(s); z \in \operatorname{supp} f_\varepsilon\}.$$

Then, by the choice of ε, $K_{s,\varepsilon} \subset \Omega$ for $0 \leqslant s \leqslant 1$. By Lemma 1.2.4, there is $v \in C_0^1(\bar{D})$ such that $\operatorname{supp} v \subset \Omega$ and

$$(\operatorname{div} v)(x) = f_\varepsilon(x - p_1) - f_\varepsilon(x - p_2). \tag{1.3.2}$$

Since ϕ is C^2 and $(\operatorname{supp} v) \cap \phi(\partial D) = \emptyset$, there is, by Lemma 1.2.2, a function $u \in C_0^1(\bar{D})$ with

$$(\operatorname{div} u)(x) = J_\phi(x) \cdot (\operatorname{div} v)(\phi(x)). \tag{1.3.3}$$

Now
$$d(\phi, D, p_1) = \int_D f_\varepsilon(\phi(x) - p_1) J_\phi(x)\, dx,$$

whence, using (1.3.2) and (1.3.3),

$$d(\phi, D, p_1) = \int_D f_\varepsilon(\phi(x) - p_2) J_\phi(x)\, dx + \int_D J_\phi(x) \cdot (\operatorname{div} v)(\phi(x))\, dx$$

$$= d(\phi, D, p_2) + \int_D (\operatorname{div} u)(x)\, dx. \tag{1.3.4}$$

(Note that we have so far in the argument restricted the size of ε three times.) By the divergence theorem,

$$\int_D (\operatorname{div} u)(x)\, dx = 0,$$

for $\operatorname{supp} u \subset D$. Hence (1.3.4) gives

$$d(\phi, D, p_1) = d(\phi, D, p_2).$$

(ii) Suppose now that ϕ is no longer C^2. There is a sequence $\{\phi_n\}$ with $\phi_n \in C^2(\bar{D})$ and $\phi_n \to \phi$ in $C^1(\bar{D})$ as $n \to \infty$. With $\gamma(s)$ as above, let $\delta = \rho(\gamma, \phi(\partial D))$; both $\phi(\partial D)$ and γ are compact, so $\delta > 0$. If $\| \phi - \phi_n \| < \frac{1}{2}\delta$, then, for $x \in \partial D$ and $0 \leqslant s \leqslant 1$,

$$|\phi_n(x)-\gamma(s)| \geqslant |\phi(x)-\gamma(s)| - |\phi(x)-\phi_n(x)|$$
$$> \tfrac{1}{2}\delta.$$

It follows that not only are p_1 and p_2 in the same component of $\mathbf{R}^n \backslash \phi(\partial D)$, but they are also in the same component of $\mathbf{R}^n \backslash \phi_n(\partial D)$. Provided we choose n sufficiently large at each stage,

$$\begin{aligned} d(\phi, D, p_1) &= d(\phi_n, D, p_1) \quad &&\text{(Theorem 1.1.5)} \\ &= d(\phi_n, D, p_2) \quad &&\text{((i) above)} \\ &= d(\phi, D, p_2) \quad &&\text{(Theorem 1.1.5).} \end{aligned}$$

The proof is now complete.

We can now remove the restriction $p \notin$ crease ϕ in our definition of degree

Definition 1.3.3 If $\phi \in C^1(\bar{D})$ and $p \notin \phi(\partial D)$ but $p \in$ crease ϕ, define $d(\phi, D, p)$ to be $d(\phi, D, q)$, where q is any point such that $q \notin$ crease ϕ and $|q-p| < \rho(p, \phi(\partial D))$.

Justification By Sard's theorem (Theorem 1.2.1) every ball $B(p, r)$ contains points $q \notin$ crease ϕ. The component of p in $\mathscr{C}\phi(\partial D)$ contains the ball $B(p, \rho(p, \phi(\partial D)))$. So, by Theorem 1.3.2, $d(\phi, D, q)$ has the same value for all $q \notin$crease ϕ satisfying $|q-p| < \rho(p, \phi(\partial D))$.

We now prove some of the properties of $d(\phi, D, p)$ for C^1 functions, regardless of whether p is a crease point or not. Theorem 1.3.5, which follows, is a partial fulfilment of the remarks made at the end of section 1.1. These properties will be used in the final stage of our definition.

Definition 1.3.4 A C^1 *homotopy* between elements ϕ and ψ of $C^1(\bar{D})$ is a function $H: \bar{D} \times [0,1] \to \mathbf{R}^n$ such that, if H_t denotes the function $x \mapsto H(x,t)$, then $H_0 = \phi$, $H_1 = \psi$, $H_t \in C^1(\bar{D})$ $(0 \leqslant t \leqslant 1)$, and $H_s \to H_t$ in $C^1(\bar{D})$ as $s \to t$ (that is, $\|H_t - H_s\|_1 \to 0$ as $s \to t$).

Theorem 1.3.5 *Let* $\phi \in C^1(\bar{D})$.

(1) $d(\phi, D, .)$ *is constant on components of* $\mathbf{R}^n \backslash \phi(\partial D)$.

(2) *If* $p \notin \phi(\partial D)$, *there is* $\varepsilon > 0$, *depending on* p *and* ϕ, *such that* $d(\phi, D, p) = d(\psi, D, p)$ *when* $\|\phi - \psi\|_1 < \varepsilon$.

(3) *Let $H(x,t)$ be a C^1 homotopy between ϕ and ψ; if $p \notin H(\partial D, t)$ for all $t \in [0,1]$, then $d(\phi, D, p) = d(\psi, D, p)$.*

Proof (1) Let Ω be a component of $\mathbf{R}^n \backslash \phi(\partial D)$, and $p_1, p_2 \in \Omega$. For $i = 1, 2$, take $q_i \notin$ crease ϕ with $|q_i - p_i| < \rho(p_i, \phi(\partial D))$; clearly $q_1, q_2 \in \Omega$. We have

$$\begin{aligned} d(\phi, D, p_1) &= d(\phi, D, q_1) \quad &&\text{(Definition 1.3.3)} \\ &= d(\phi, D, q_2) \quad &&\text{(Theorem 1.3.2)} \\ &= d(\phi, D, p_2) \quad &&\text{(Definition 1.3.3).} \end{aligned}$$

(2) Let $\eta = \rho(p, \phi(\partial D))$. Choose q so that $|q - p| < \frac{1}{2}\eta$ and $q \notin$ crease ϕ (such a choice is possible, by Sard's theorem); choose $\varepsilon < \frac{1}{2}\eta$ to be such that $\| \phi - \psi \|_1 < \varepsilon$ implies that $q \notin$ crease ψ and $d(\phi, D, q) = d(\psi, D, q)$ (this choice is possible by Theorem 1.1.5). If $x \in \partial D$, then

$$|p - \psi(x)| \geqslant |p - \phi(x)| - |\phi(x) - \psi(x)| > \tfrac{1}{2}\eta.$$

Hence p and q are in the same component of $\mathbf{R}^n \backslash \psi(\partial D)$ as well as in the same component of $\mathbf{R}^n \backslash \phi(\partial D)$. It follows that

$$\begin{aligned} d(\phi, D, p) &= d(\phi, D, q) \quad &&\text{(part (1) above)} \\ &= d(\psi, D, q) \quad &&\text{(the choice of } \varepsilon) \\ &= d(\psi, D, p) \quad &&\text{(part (1) again).} \end{aligned}$$

(3) $d(H_t, D, p)$ is defined for all $t \in [0,1]$, and by (2) it is a continuous function from $[0,1]$ into $\mathbf{R}$. Since d is integer-valued, it must be independent of t in the range $[0,1]$.

Remark The intuitive idea behind (3) is that the degree can change only if the deformation which the homotopy represents analytically pulls a p-point across the boundary ∂D.

1.4 Continuous functions

In this section we come to the final stage of the definition of degree in finite dimensional spaces. We suppose only that $\phi \in C(\bar{D})$; the degree of ϕ is then the degree of a sufficiently good C^1 approximation to ϕ. That a definition of degree is possible for non-differentiable functions demonstrates the topological nature of the concept–in these terms the analytic formulation we have pursued is merely a means of calculation.

Definition 1.4.1 Suppose that $\phi \in C(\bar{D})$ and $p \notin \phi(\partial D)$. Define $d(\phi, D, p)$ to be $d(\psi, D, p)$, where ψ is any function in $C^1(\bar{D})$ satisfying

$$|\phi(x) - \psi(x)| < \rho(p, \phi(\partial D)) \quad (x \in \bar{D}). \tag{1.4.1}$$

Remark The bound in (1.4.1) is relatively unimportant; conceptually it is 'if $\| \phi - \psi \|$ is small enough'.

Justification In every neighbourhood of ϕ in $C(\bar{D})$ there are C^1 functions. Let $\rho(p, \phi(\partial D)) = \eta$, and suppose that, for $i = 1, 2$, $\psi_i \in C^1(\bar{D})$ and $\| \phi - \psi_i \| < \eta$. Consider the C^1 homotopy

$$h_t(x) = t\psi_1(x) + (1-t)\psi_2(x) \quad (x \in \bar{D}, 0 \leqslant t \leqslant 1).$$

Now $|h_t(x) - \phi(x)| = |t(\psi_1(x) - \phi(x)) + (1-t)(\psi_2(x) - \phi(x))|$

$$< t\eta + (1-t)\eta = \eta.$$

So, if $x \in \partial D$,

$$\begin{aligned} |p - h_t(x)| &\geqslant |p - \phi(x)| - |\phi(x) - h_t(x)| \\ &> 0. \end{aligned}$$

Thus $p \notin h_t(\partial D)$ for $0 \leqslant t \leqslant 1$, whence $d(\psi_1, D, p) = d(\psi_2, D, p)$, by Theorem 1.3.5(3). We conclude that $d(\psi, D, p)$ is the same for all $\psi \in C^1(\bar{D})$ within η of ϕ; Definition 1.4.1 is therefore meaningful.

The usual procedure in the proofs of degree-theoretic results is to first prove the result for the 'nice' case of C^1 functions and non-crease points p, and then so seek to prove the general result by a process of approximation (an observation that will be amply illustrated in the sequel!). The next theorem is of considerable help in this scheme, for it tells us in effect that we can approximate to ϕ and p simultaneously.

Theorem 1.4.2 *In Definition* 1.4.1, *the function ψ can be so chosen that $p \notin$* crease *ψ.*

Proof Let $\rho(p, \phi(\partial D)) = \eta$. Take $\chi \in C^1(\bar{D})$ such that $\|\phi - \chi\| < \frac{1}{2}\eta$; then $\rho(p, \chi(\partial D)) > \frac{1}{2}\eta$. Choose q so that $|q - p| < \frac{1}{2}\eta$ and $q \notin$ crease χ. Let $\psi(x) = \chi(x) + p - q$; then

$$\|\phi - \psi\| \leqslant \|\phi - \chi\| + |p - q| < \eta.$$

Now $\psi(x) = p$ if and only if $\chi(x) = q$, and for such x, $J_\psi(x) = J_\chi(x) \neq 0$; hence $p \notin$ crease ψ. So ψ satisfies (1.4.1) and does not have p as a critical value.

We now look at some simple illustrations in the hope that they will serve to confirm the reader's impressions. They were referred to at the end of Section 1.1.

Illustration 1 Let $n = 1$, $D = (a, b)$; consider a continuous function f such as that shown in Figure 1.4. The degree $d(f, D, p)$ is defined unless $p = f(a)$ or $p = f(b)$. The critical values of f are p_1, p_2, p_3 and p_4. For all other values of $p, d(f, D, p)$ can be calculated from Definition 1.1.3. If $p \notin [f(a), f(b)]$, then $d(f, D, p) = 0$. If $f(a) < f(b)$ and $p \in (f(a), f(b))$, we have $d(f, D, p) = +1$; should we have

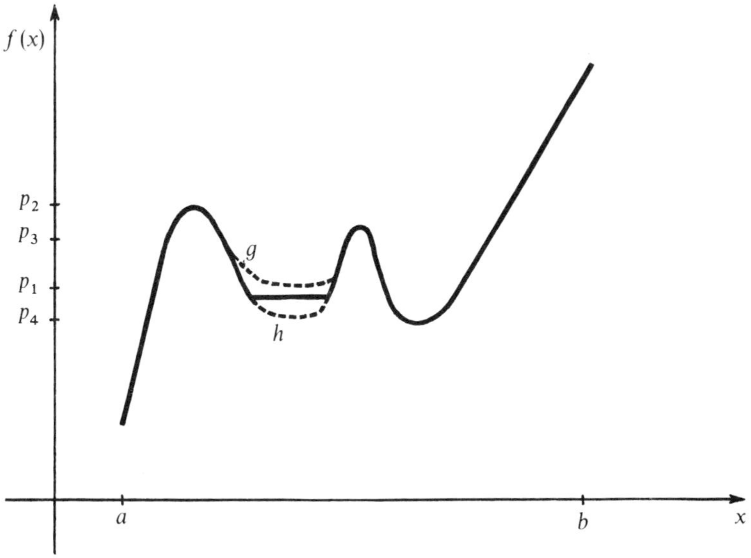

Figure 1.4. The functions of Illustration 1

$f(a) > f(b)$ and $p \in (f(b), f(a))$, then $d(f, D, p) = -1$. The degree $d(f, D, p_i)$ ($i = 2$, 3 or 4) is found by considering $d(f, D, q)$, where q is a non-crease point near p_i; thus $d(f, D, p_i) = +1, 0$ or -1 according as $f(a) < p_i < f(b)$, $p_i \notin [f(a), f(b)]$ or $f(b) < p_i < f(a)$. As for $d(f, D, p_1)$, we use C^1 functions such as g and h (shown in broken lines in Figure 1.4) for which p_1 is not a crease point. In the configuration of the figure, $d(g, D, p_1) = d(h, D, p_1) = +1$, whence $d(f, D, p_1) = 1$. Note that $d(f, D, p)$ depends only on the values of f on $\partial D = \{a, b\}$; we shall later prove that this is generally true. It is instructive to deform the graph of f into other shapes and to observe the consequent changes in the degree.

Illustration 2 Suppose that D is a bounded domain in $\mathbf{C}$, which we identify with $\mathbf{R}^2$. Consider $\phi = \phi_1 + i\phi_2$, a holomorphic function defined on $\bar{D}$. If z_0 is a zero of the function $\phi - p$ of multiplicity m, say, we can write

$$\phi(z) - p = (z - z_0)^m g(z)$$

in a neighbourhood U of z_0, where $g(z_0) \neq 0$. By the Cauchy–Riemann equations, $J_\phi = |\phi'|^2 \geqslant 0$. If U is small enough, J_ϕ is non-zero in $U' \equiv U \setminus \{z_0\}$; moreover, U' contains exactly m solutions of $\phi(z) = p + \delta$, none of them multiple, if δ is small enough and non-zero. Similar remarks apply to each of the finite number of p-points of ϕ contained in D. So $d(\phi, D, p + \delta)$ is simply the number of solutions of $\phi(z) = p + \delta$ in D. But $d(\phi, D, p + \delta) = d(\phi, D, p)$ for small enough δ; therefore

$$d(\phi, D, p) = \text{number of } p\text{-points of } \phi \text{ in } D, \text{ counting multiplicity.}$$

The right hand side of the last equation is, of course,

$$\frac{1}{2\pi i} \int_{\partial D} \frac{\phi'(z)}{\phi(z) - p} \, dz$$

provided that ∂D is a sufficiently 'nice' curve.

Remark Illustration 2 effectively shows that degree can be regarded as a generalised winding number. We again note that $d(\phi, D, p)$ depends only on the values of ϕ on the boundary ∂D.

Illustration 3 In the third illustration we look at some mappings of $\mathbf{R}^2$ into itself, and seek their degrees at 0 relative to the unit disc B.

(i) $$\phi(x, y) = (|y| - x, x^2 + 2x + y^2),$$

(ii) $$\psi(x, y) = (y - x^3, y),$$

(iii) $$\chi(x, y) = (e^x - 1, y^2).$$

All three of these mappings have a unique zero, at 0; 0 is a critical point of ψ and χ; ϕ is not C^1.

(i) The mapping $\phi_\varepsilon : (x, y) \mapsto (|y| - x + \varepsilon, x^2 + 2x + y^2)$ has no zeros if $0 < \varepsilon < 1$, whence $d(\phi, B, 0) = 0$.

(ii) To calculate $d(\psi, B, 0)$, we consider the mappings

$$\psi_\varepsilon : (x, y) \mapsto (y - x^3 + \varepsilon^2 x, y) \quad (\varepsilon \to 0).$$

ψ_ε has three zeros in B: $(0, 0)$, $(\varepsilon, 0)$ and $(-\varepsilon, 0)$. The Jacobian J_{ψ_ε} is positive at $(0, 0)$ and negative at both $(\varepsilon, 0)$ and $(-\varepsilon, 0)$. Therefore $d(\psi, B, 0) = d(\psi_\varepsilon, B, 0) = -1$.

(iii) For χ, consider, say, $\chi_\varepsilon : (x, y) \mapsto (e^x - 1, y^2 - \varepsilon)$. We see that $\chi_\varepsilon^{-1}(0) = \{(0, \varepsilon), \quad (0, -\varepsilon)\}$; $J_{\chi_\varepsilon}(0, \varepsilon) = 2\varepsilon = -J_{\chi_\varepsilon}(0, -\varepsilon)$. Hence $d(\chi, D, 0) = d(\chi_\varepsilon, D, 0) = 0$.

The reader will be able to construct an unending variety of examples such as these; a geometric interpretation (and the drawing of diagrams) is useful. The degree of polynomial mappings defined on the plane are considered in some detail in Cronin (1964).

When we come to define degree in infinite dimensional spaces in Chapter 4, we shall need to apply the theory developed in this chapter and the next in finite dimensional spaces other than $\mathbf{R}^n$. Suppose that X is a real normed space of dimension n; X can be identified with $\mathbf{R}^n$ once a basis has been chosen. Thus the degree of mappings from X can be defined provided we can show that the degree we have defined in $\mathbf{R}^n$ is independent of the basis chosen.

Theorem 1.4.3 *If $D \subset \mathbf{R}^n$, $\phi \in C(\bar{D})$ and $p \notin \phi(\partial D)$, then $d(\phi, D, p)$ is invariant under a non-singular C^1 change of co-ordinates.*

Proof Let the change of co-ordinates be given by the C^1 injective function $\psi : \psi^{-1}(\bar{D}) \to \bar{D}$, whose Jacobian J_ψ is nowhere zero. Let the new co-ordinates be denoted by y and the new form of ϕ by ϕ^*. Then $\phi^*(y) = \psi^{-1} \circ \phi \circ \psi(y)$. J_ψ and $J_{\psi^{-1}}$ are both nowhere zero. If $\phi \in C^1(\bar{D})$, it follows at once that sign $J_{\phi^*}(y) = \text{sign } J_\phi(x)$; thus, if $p \notin \text{crease } \phi$, $d(\phi, D, p)$ is the same whether calculated in terms of the x co-ordinates or the y co-ordinates. By considering suitable approximations to ϕ and p, it is seen that $d(\phi, D, p)$ is also unchanged by the change of co-ordinates when $p \in \text{crease } \phi$ and $\phi \in C(\bar{D}) \setminus C^1(\bar{D})$.

Corollary 1.4.4 *The theory of Chapter* 1 *is valid when* $\mathbf{R}^n$ *is replaced by another normed space of dimension n.*

If E_1 and E_2 are two normed spaces of the same finite dimension, we can reasonably expect to be able to apply the theory to mappings $\phi : D \to E_2$, where $D \subset E_1$. We obtain a uniquely defined degree by following the previous definitions provided that E_1 and E_2 have specified orientations (that is, are 'oriented').

Corollary 1.4.5 *The theory of Chapter* 1 *is applicable to mappings between two oriented normed spaces of the same finite dimension.*

2
Properties of degree in finite dimensional spaces

In this chapter we prove those properties of $d(\phi, D, p)$ that are most useful in the sequel and in applications.

2.1 Changes in ϕ and p

The first result is the basis of many of the applications of degree theory in analysis.

Theorem 2.1.1 *Suppose that $\phi \in C(\bar{D})$. If $d(\phi, D, p)$ is defined and non-zero, then there is $q \in D$ such that $\phi(q) = p$.*

Proof If $p \notin \phi(\bar{D})$, take $\psi \in C^1(\bar{D})$ such that $\|\phi - \psi\| < \rho(p, \phi(\bar{D}))$. Then $p \notin \psi(\bar{D})$, whence $d(\psi, D, p) = 0$ (by the definition). So we have $d(\phi, D, p) = 0$, using Definition 1.4.1. Hence $p \in \phi(D)$ if $d(\phi, D, p) \neq 0$.

Recall that if X and Y are topological spaces, two continuous functions f and g are said to be homotopic ($f \sim g$) if there is a continuous function

$$H : [0, 1] \times X \to Y$$

such that

$$H(0, x) = f(x), \quad H(1, x) = g(x) \quad (x \in X).$$

Intuitively, this means that the graph of f is continuously deformed into that of g.

Theorem 2.1.2 (1) *Suppose that $\phi \in C(\bar{D})$ and $p \notin \phi(\partial D)$. If $\|\psi - \phi\| < \rho(p, \phi(\partial D))$, then $d(\psi, D, p)$ is defined and equals $d(\phi, D, p)$.*

(2) *If $H(t, x) \equiv h_t(x)$ is a homotopy and $p \notin h_t(\partial D)$ for $0 \leqslant t \leqslant 1$, then $d(h_t, D, p)$ is independent of $t \in [0, 1]$.*

Proof (1) Since $\|\psi-\phi\|<\rho(p,\phi(\partial D))$ and $p\notin\phi(\partial D)$, we have that $p\notin\psi(\partial D)$, so that $d(\psi,D,p)$ is defined. Take $\chi\in C^1(\bar{D})$ such that

$$\|\chi-\psi\|+\|\psi-\phi\|<\rho(p,\phi(\partial D));$$

such a choice is certainly possible. Then, clearly, $\|\chi-\phi\|<\rho(p,\phi(\partial D))$; so, by Definition 1.4.1, $d(\phi,D,p)=d(\chi,D,p)$. Since

$$\rho(p,\phi(\partial D))\leqslant\rho(p,\psi(\partial D))+\|\phi-\psi\|,$$

we also have $\|\chi-\psi\|<\rho(p,\psi(\partial D))$. Thus, again by Definition 1.4.1, $d(\psi,D,p)=d(\chi,D,p)$. We conclude that

$$d(\psi,D,p)=d(\phi,D,p)\quad\text{if}\quad\|\psi-\phi\|<\rho(p,\phi(\partial D)).$$

(2) The hypotheses ensure that $d(h_t,D,p)$ is defined for all $t\in[0,1]$. By (1), $t\mapsto d(h_t,D,p)$ is a continuous function from $[0,1]$ into the integers; such a function is constant, whence the result.

Remarks Part (1) of the theorem is often interpreted as 'if ψ is near enough to ϕ, then the degrees of ψ and ϕ are equal'; the estimate given, though, is frequently useful. Another interpretation is that $d(\phi,D,p)$ is continuous in ϕ. The next result will tell us that $d(\phi,D,p)$ is also continuous in p. Part (2) of Theorem 2.1.2 is known as 'homotopy invariance'. Heuristically, provided p-points do not occur on the boundary ∂D during the deformation H, then p-points appear and disappear in pairs contributing zero to the degree.

Theorem 2.1.3 *$d(\phi,D,.)$ is constant on components of $\mathbf{R}^n\backslash\phi(\partial D)$.*

Proof Let $p_1,p_2\in\Omega$, a component of $\mathbf{R}^n\backslash\phi(\partial D)$. As we have seen before there exists a path $X(s)$ in Ω joining p_1 and p_2. Take $\psi\in C^1(\bar{D})$ with $\|\phi-\psi\|<\rho(X[0,1],\phi(\partial D))$. Then $d(\phi,D,p_i)=d(\psi,D,p_i)$ $(i=1,2)$, and p_1,p_2 are in the same component of $\mathbf{R}^n\backslash\psi(\partial D)$. The result now follows from the corresponding theorem for C^1 functions (Theorem 1.3.5).

Notation If A is a connected subset of $\mathbf{R}^n\backslash\phi(\partial D)$, we write $d(\phi,D,A)$ for $d(\phi,D,p)$ $(p\in A)$.

Next we come to a simple but important fact, namely that $d(\phi, D, p)$ depends only on the values ϕ takes on ∂D. This further illustrates the 'winding number' nature of degree.

Theorem 2.1.4 (Boundary value dependence) *If $\phi, \psi \in C(\bar{D})$ and $\phi = \psi$ on ∂D, then $d(\phi, D, p) = d(\psi, D, p)$ (provided that $p \notin \phi(\partial D)$).*

Proof Consider the homotopy

$$H(t, x) = t\phi(x) + (1-t)\psi(x) \quad (x \in \bar{D},\ 0 \leqslant t \leqslant 1).$$

For $x \in \partial D$, $H(t, x) = \phi(x)$, so certainly $p \notin H(t, \partial D)$ $(0 \leqslant t \leqslant 1)$. Therefore, by homotopy invariance, $d(\phi, D, p) = d(\psi, D, p)$.

Theorem 2.1.5 (Poincaré–Bohl theorem) *Let $\phi, \psi \in C(\bar{D})$, and suppose that for all $x \in \partial D$, the line segment $[\phi(x), \psi(x)]$ does not contain p. Then $d(\phi, D, p) = d(\psi, D, p)$.*

Proof Again we use the homotopy

$$H(t, x) = t\phi(x) + (1-t)\psi(x) \quad (x \in \bar{D},\ 0 \leqslant t \leqslant 1).$$

The condition of the theorem means that the vectors $\phi(x) - p$, $\psi(x) - p$ are never in opposition on ∂D. Thus

$$t(\phi(x) - p) + (1-t)(\psi(x) - p) \neq 0 \quad (x \in \partial D,\ 0 \leqslant t \leqslant 1).$$

This means that

$$H(t, x) \neq p \quad (x \in \partial D,\ 0 \leqslant t \leqslant 1),$$

from which the result follows by homotopy invariance.

Theorem 2.1.6 *Suppose that $\phi \in C(\bar{D})$ and $p \notin \phi(\partial D)$. Then, for all $q \in \mathbf{R}^n$,*

$$d(\phi, D, p) = d(\phi - q, D, p - q)$$

(where $\phi - q$ denotes the mapping $x \mapsto \phi(x) - q$).

Proof Consider the function $\Phi: [0, 1] \to \mathbf{Z}$ given by

$$\Phi(t) = d(\phi - tq, D, p - tq).$$

Φ is defined for $t \in [0, 1]$ because $p - tq \in (\phi - tq)(\partial D)$ implies that $p \in \phi(\partial D)$. By Theorems 2.1.2 and 2.1.3, Φ is continuous; being integer-valued, it is constant. The result follows.

From Theorem 2.1.6, we can deduce the following form of homotopy invariance.

Theorem 2.1.7 *Suppose that* $h_t (0 \leqslant t \leqslant 1)$ *is a homotopy in* $C(\bar{D})$ *and* p_t *is a continuous path in* $\mathbf{R}^n$. *If* $p_t \notin h_t(\partial D)$ $(0 \leqslant t \leqslant 1)$, *then* $d(h_t, D, p_t)$ *is independent of* $t \in [0, 1]$.

Proof Consider the homotopy

$$k_t(x) = h_t(x) - p_t \quad (0 \leqslant t \leqslant 1,\ x \in \bar{D}).$$

By Theorem 2.1.6, $d(k_t, D, 0) = d(h_t, D, p_t)$; by Theorem 2.1.2(2), $d(k_t, D, 0)$ is independent of $t \in [0, 1]$, whence the same is true of $d(h_t, D, p_t)$.

2.2 Changes in the domain D

We have seen how $d(\phi, D, p)$ changes with ϕ and p; it is now the turn of D to be varied. First we consider subdivisions of D.

Theorem 2.2.1 *Suppose that* $p \notin \phi(\partial D)$ *and* $\phi \in C(\bar{D})$.

(1) *If* D *is the disjoint union of open sets* D_i $(i = 1, 2, \ldots)$, *then*

$$d(\phi, D, p) = \sum_i d(\phi, D_i, p).$$

(2) *If* $K \subset \bar{D}$ *is closed and* $p \notin \phi(K)$, *then*

$$d(\phi, D, p) = d(\phi, D \backslash K, p).$$

Proof (1) To start with we show that $\partial D_i \subset \partial D$ for all i. It is clear that $\partial D_i \subset \bar{D}_i \subset \bar{D}$. If there is $y \in \partial D_i$ with $y \notin \partial D$, we must have $y \in D$. Hence $y \in D_j$ for some $j \neq i$. But D_j is open, so that there is a neighbourhood U of y contained in D_j; because the D_k are disjoint, $U \cap D_i = \emptyset$. This contradicts the hypothesis that $y \in \partial D_i$. We conclude that $\partial D_i \subset \partial D$ (for all i).

Take $\psi \in C^1(\bar{D})$ such that $p \notin$ crease ψ and

$$\| \phi - \psi \| < \rho(p, \phi(\partial D)).$$

Since $\partial D_i \subset \partial D, p \notin \psi(\partial D_i)$ and

$$|\phi(x) - \psi(x)| < \rho(p, \phi(\partial D_i)) \quad (x \in \bar{D}_i).$$

Hence $d(\phi, D_i, p) = d(\psi, D_i, p)$ for all i. Thus

$$d(\phi, D, p) = d(\psi, D, p) = \sum_{x \in \psi^{-1}(p)} \operatorname{sign} J_\psi(x)$$
$$= \sum_j \sum_{x \in \psi^{-1}(p) \cap D_j} \operatorname{sign} J_\psi(x)$$
$$= \sum_j d(\psi, D_j, p)$$
$$= \sum_j d(\phi, D_j, p).$$

Note that these summations are all finite – for $\psi^{-1}(p)$ is a finite set.

(2) Choose $\psi \in C^1(\bar{D})$ such that $p \notin$ crease ψ, $\| \phi - \psi \| < \rho(p, \phi(\partial D))$, and furthermore $\| \phi - \psi \| < \rho(p, \phi(K))$–since K is compact, $\rho(p, \phi(K)) > 0$. Then $p \notin \psi(K)$, and we have

$$d(\phi, D, p) = d(\psi, D, p)$$
$$= \sum_{x \in \psi^{-1}(p) \cap D} \operatorname{sign} J_\psi(x)$$
$$= \sum_{x \in \psi^{-1}(p) \cap (D \backslash K)} \operatorname{sign} J_\psi(x) \quad (\text{for } \psi^{-1}(p) \cap K = \emptyset)$$
$$= d(\psi, D \backslash K, p).$$

Now $\| \phi - \psi \| < \rho(p, \phi(\partial(D \backslash K))$ by the choice of ψ, whence

$$d(\psi, D \backslash K, p) = d(\phi, D \backslash K, p).$$

The result is therefore proved.

Remark Property (1) of the theorem is termed '*decomposition of domain*' and (2) is called the '*excision*' property.

Part (2) of Theorem 2.2.1 enables us to introduce the idea of the index of an isolated solution of $\phi(x) = p$. For $n = 2$ it is the same as the so-called Poincaré index which is frequently used in the qualitative theory of differential equations (and often introduced with insufficient care).

Suppose that $\phi \in C(\bar{D})$ and that x_0 is an isolated p-point of ϕ in D. Let $\mathcal{U}$ be the collection of all open neighbourhoods of x_0 not containing another p-point of ϕ. If $U_1, U_2 \in \mathcal{U}$, then $U_1 \cup U_2 \in \mathcal{U}$, and the excision property implies that $d(\phi, U_1, p) = d(\phi, U_1 \cup U_2, p) = d(\phi, U_2, p)$. Therefore $d(\phi, U, p)$ is the same for all $U \in \mathscr{U}$.

Definition 2.2.2 The *index* $i(\phi, x_0, p)$ of x_0 is the common value of $d(\phi, U, p)$ for $U \in \mathscr{U}$.

The following results are of frequent use in bifurcation theory.

Theorem 2.2.3 (1) *If $\phi \in C(\bar{D})$, $p \notin \phi(\partial D)$ and $\phi^{-1}(p)$ is finite, then*

$$d(\phi, D, p) = \sum_{a \in \phi^{-1}(p)} i(\phi, a, p).$$

(2) *If $\phi \in C^1(\bar{D})$, $a \in \phi^{-1}(p)$ and $J_\phi(a) \neq 0$, then $i(\phi, a, p) = (-1)^\nu$, where ν is the number of real negative eigenvalues of $\phi'(a)$ (counting algebraic multiplicity).*

Proof (1) Let $\phi^{-1}(p) = \{a_1, \ldots, a_k\}$, and suppose N_i $(i = 1, \ldots, k)$ are disjoint neighbourhoods of a_i such that $d(\phi, N_i, p) = i(\phi, a_i, p)$. Using the excision property and that of decomposition of domain, the result follows easily.

(2) The linear operator $\phi'(a)$ is invertible. If $\lambda_1, \ldots, \lambda_n$ are the eigenvalues of $\phi'(a)$ (not necessarily distinct), then $J_\phi(a) = \lambda_1 \ldots \lambda_n$. Complex eigenvalues occur in conjugate pairs, so

$$\operatorname{sign} J_\phi(a) = (-1)^\nu.$$

But $i(\phi, a, p)$ is defined as sign $J_\phi(a)$.

We conclude this section with a form of homotopy invariance which is sometimes useful. The proof we leave to the reader; it combines the results of Theorem 2.2.1 with Theorem 2.1.2(2).

Theorem 2.2.4 *Suppose that D_* is a bounded, open subset of $[0, 1] \times \mathbf{R}^n$ and that $\phi: D_* \to \mathbf{R}^n$ is continuous. Let ϕ_t denote the mapping $x \mapsto \phi(t, x)$, and let $D_t = \{x; (t, x) \in D_*\} \subset \mathbf{R}^n$. If $p \notin \phi_t(\partial D_t)$ for $0 \leqslant t \leqslant 1$, then $d(\phi_t, D_t, p)$ is independent of $t \in [0, 1]$.*

Remark If the hypotheses of Theorem 2.2.4 are satisfied and p_t is a continuous path in $\mathbf{R}^n$ such that $p_t \notin \phi_t(\partial D_t)$ for $t \in [0, 1]$, then $d(\phi_t, D_t, p_t)$ is independent of t.

2.3 The multiplication theorem

Of those properties of degree in $\mathbf{R}^n$ which we wish to present, the most difficult to prove is the following. It is concerned with the composition of functions (see Leray (1935, 1950)).

Theorem 2.3.1 (Multiplication theorem) *Let $\phi \in C(\bar{D})$ and M be a bounded, open set containing $\phi(\bar{D})$. Let $\Delta = M \backslash \phi(\partial D)$, and suppose that the components of Δ are $\Delta_i\,(i = 1, 2, \ldots)$. If $\psi \in C(\bar{M})$ and $p \notin \psi(\phi(\partial D)) \cup \psi(\partial M)$, then*

$$d(\psi \circ \phi, D, p) = \sum_j d(\psi, \Delta_j, p) d(\phi, D, \Delta_j). \tag{2.3.1}$$

Proof Note first that Δ can have at most countably many components; to see this we simply observe that every component of an open set such as Δ contains a point with rational co-ordinates. The summation in (2.3.1) is finite – for $\psi^{-1}(p)$ is compact, so that, being covered by the disjoint, open sets Δ_j, it meets only a finite number of the Δ_j.

Figure 2.1 will help to disentangle the complications; with the same aim we subdivide the proof.

(*a*) First, $\partial\Delta \subset \phi(\partial D) \cup \partial M$, so that $p \notin \psi(\partial\Delta)$. Hence $d(\psi, \Delta_i, p)$ is defined for all i.

(*b*) Let $W_k = \{y \in M; d(\phi, D, y) = k\}$. Clearly

$$W_k = \bigcup \{\Delta_i; d(\phi, D, \Delta_i) = k\}. \tag{2.3.2}$$

As we have already noted, the only non-zero summands in (2.3.1) are those corresponding to the finite number of Δ_j which cover $R = \psi^{-1}(p)$. Similarly R is covered by finitely many of the W_k. We have

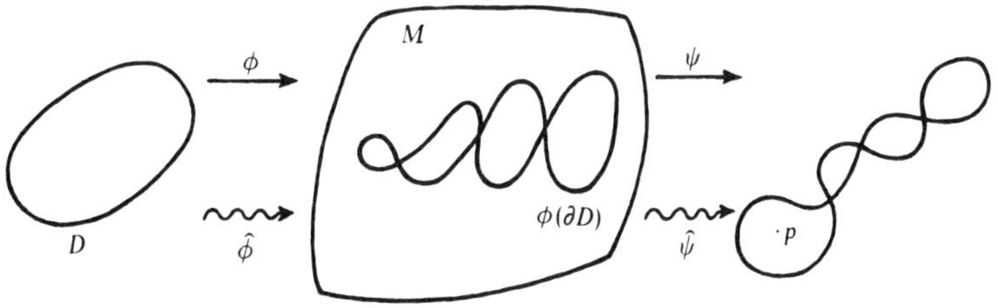

Figure 2.1. The multiplication theorem

$$\sum_j d(\psi,\Delta_j,p)d(\phi,D,\Delta_j)=\sum_k\sum_{\{j;\Delta_j\subset W_k\}} d(\psi,\Delta_j,p)d(\phi,D,\Delta_j)$$

$$=\sum_k kd(\psi,W_k,p) \tag{2.3.3}$$

(the second equality by the property of domain decomposition).

(*c*) Because $p\notin\psi\circ\phi(\partial D)$, R and $\phi(\partial D)$ are disjoint (compact) sets. Take $\hat{\phi}\in C^1(\bar{D})$ such that

$$|\phi(x)-\hat{\phi}(x)|<\rho(R,\phi(\partial D)) \quad (x\in\bar{D}). \tag{2.3.4}$$

Then
$$d(\phi,D,y)=d(\hat{\phi},D,y) \quad (y\in R). \tag{2.3.5}$$

Let
$$\hat{W}_k=\{y\in M; d(\hat{\phi},D,y)=k\}.$$

Now $d(\phi,D,y)$ is unchanged under small changes in y, and W_k is an open set. If $y\in R$ and $y\notin W_k$, then, since $R\cap\phi(\partial D)=\emptyset$, $d(\phi,D,y)$ is defined, and $y\in\operatorname{int}(\mathscr{C}W_k)$. Therefore $R\cap\partial W_k=\emptyset$. By a similar argument (and using (2.3.5)) $R\cap\partial\hat{W}_k=\emptyset$ also. If $z\in\psi^{-1}(p)\cap W_k$, then $d(\phi,D,z)=k$, whence $d(\hat{\phi},D,z)=k$ (by (2.3.5)); this means that $z\in\psi^{-1}(p)\cap\hat{W}_k$. Similarly $z\in\psi^{-1}(p)\cap\hat{W}_k$ implies that $z\in\psi^{-1}(p)\cap W_k$. Therefore

$$\psi^{-1}(p)\cap W_k=\psi^{-1}(p)\cap\hat{W}_k,$$

that is, $\psi^{-1}(p)\cap(W_k\cup\hat{W}_k)\subset W_k\cap\hat{W}_k$. Then, applying the excision property to $W_k\cup\hat{W}_k$, we have $d(\psi,W_k\cup\hat{W}_k,p)=d(\psi,W_k\cap\hat{W}_k,p)=d(\psi,W_k,p)$, and similarly $d(\psi,W_k\cup\hat{W}_k,p)=d(\psi,\hat{W}_k,p)$. Thus

$$d(\psi,W_k,\ p)=d(\psi,\hat{W}_k,p). \tag{2.3.6}$$

(*d*) Choose $\hat{\psi}\in C^1(\bar{M})$ such that $p\notin\text{crease}\,(\hat{\psi}\circ\hat{\phi})$ and

$$|\hat{\psi}(x)-\psi(x)|<\rho(p,\psi(\partial M)\cup\psi\circ\hat{\phi}(\partial D)) \quad (x\in\bar{M}). \tag{2.3.7}$$

(This can be done by the kind of argument used in Theorem 1.4.2.) In particular, $p\notin\hat{\psi}\circ\hat{\phi}(\partial D)$. As we have seen, $d(\hat{\phi},D,y)$ is undefined for $y\in\partial\hat{W}_k$; this means that $\partial\hat{W}_k\subset\hat{\phi}(\partial D)\cup\partial M$. Hence $|\hat{\psi}(x)-\psi(x)|<\rho(p,\psi(\partial\hat{W}_k))$ for $x\in\bar{M}$, so that $p\notin\hat{\psi}(\partial\hat{W}_k)$ and

$$d(\psi,\hat{W}_k,p)=d(\hat{\psi},\hat{W}_k,p). \tag{2.3.8}$$

(*e*) If we can equate the degrees of $\psi\circ\phi$ and $\hat{\psi}\circ\hat{\phi}$, the equality (2.3.1) is reduced to the case of C^1 functions. Let

$$\phi_t(x) = (1-t)\phi(x) + t\hat{\phi}(x) \quad (x \in \bar{D}, 0 \leqslant t \leqslant 1),$$

and consider the homotopy $\psi \circ \phi_t$. By (2.3.4), $\rho(R, \phi_t(\partial D)) > 0$ for all $t \in [0, 1]$, whence $p \notin \psi \circ \phi_t(\partial D)$, and by homotopy invariance

$$d(\psi \circ \phi, D, p) = d(\psi \circ \hat{\phi}, D, p). \tag{2.3.9}$$

Consider also the homotopy

$$\psi_t \circ \hat{\phi} = ((1-t)\psi + t\hat{\psi}) \circ \hat{\phi} \quad (0 \leqslant t \leqslant 1).$$

By an argument similar to the above, $p \notin \psi_t \circ \hat{\phi}(\partial D)$ and

$$d(\psi \circ \hat{\phi}, D, p) = d(\hat{\psi} \circ \hat{\phi}, D, p). \tag{2.3.10}$$

(*f*) Putting (2.3.3), (2.3.6), (2.3.8), (2.3.9) and (2.3.10) together, we have completed the proof if we can show that

$$\sum_k k d(\hat{\psi}, \hat{W}_k, p) = d(\hat{\psi} \circ \hat{\phi}, D, p). \tag{2.3.11}$$

Recall that $\hat{\phi} \in C^1(\bar{D})$, $\hat{\psi} \in C^1(\bar{M})$, $p \notin \hat{\psi} \circ \hat{\phi}(\partial D) \cup \hat{\psi}(\partial \hat{W}_k)$ and $p \notin$ crease$(\hat{\psi} \circ \hat{\phi})$. The problem has been reduced to a computation.

It is convenient in the calculation to remove the ^ from these expressions. Now by a well-known property of Jacobians

$$d(\psi \circ \phi, D, p) = \sum_{x \in \phi^{-1}(\psi^{-1}(p))} \operatorname{sign}[J_\psi(\phi(x)) J_\phi(x)],$$

where the summation is again finite. So

$$\begin{aligned} d(\psi \circ \phi, D, p) &= \sum_{y \in \psi^{-1}(p)} \operatorname{sign} J_\psi(y) \left(\sum_{x \in \phi^{-1}(y)} \operatorname{sign} J_\phi(x) \right) \\ &= \sum_{y \in \psi^{-1}(p)} \operatorname{sign} J_\psi(y) d(\phi, D, y) \\ &= \sum_k \sum_{y \in \psi^{-1}(p) \cap W_k} k \operatorname{sign} J_\psi(y) \\ &= \sum_k k d(\phi, W_k, p). \end{aligned}$$

This was the required equality (2.3.11).

Remark It is well to emphasise that M can be chosen to be any convenient bounded, open set containing $\phi(\bar{D})$. It must, of course, be chosen so that $\psi^{-1}(p)$ does not meet ∂M.

2.4 Mappings defined on manifolds

This book is mainly concerned with the degree of mappings defined on subsets of linear spaces, of finite and of infinite dimension. In this section we briefly remark on how the ideas encountered relate to mappings between manifolds. This is not an integral part of the development of the book – it is an aside, and no subsequent parts are dependent on it. We do not define the new terms that arise, neither do we give proofs. Details of this and related material may be found in Milnor (1965) and Pontryagin (1959).

Let M and N be oriented, smooth manifolds of dimension n and let $\phi: M \to N$ be continuous. If $p \in N$ and y is an isolated p-point of ϕ, then, given local co-ordinates at y and $\phi(y)$, the index $i(\phi, y, p)$ can be calculated. Since both manifolds are oriented, Theorem 1.4.3 shows that the result of this calculation is independent of the particular local co-ordinates; thus $i(\phi, y, p)$ is well defined. If M is a compact manifold, then the degree $d(\phi, M, p)$ can be defined exactly as in Chapter 1 provided that p is not the image of a boundary point of M under ϕ and p is not a critical value of ϕ. We still have Sard's theorem, so that this last restriction can be removed as before. If M is a manifold without boundary, then $d(\phi, M, p)$ is defined for all p. It is not difficult to see that, again, $d(\phi, M, p) = d(\psi, M, p)$ if ϕ and ψ are homotopic mappings.

Suppose now that N is connected; then $d(\phi, M, p)$ is independent of p. We can therefore define $d(\phi)$, the degree of ϕ. We have already said that $d(\phi) = d(\psi)$ if ϕ and ψ are homotopic; conversely, we have a striking result, proved by Hopf (1926*a*).

Theorem 2.4.1 *Let $C(M)$ be the set of continuous mappings from the connected, oriented, smooth n-manifold M into the oriented sphere S^n. If $\phi, \psi \in C(M)$, then ϕ and ψ are homotopic if and only if $d(\phi) = d(\psi)$. Moreover, there exist maps of any given degree.*

As an example, take $M = S^n$; d is a mapping from the homotopy group $\pi_n(S^n)$ into the integers. Theorem 2.4.1 tells us that this mapping is one to one and onto. Thus $\pi_n(S^n) \cong \mathbf{Z}$.

Let us now consider vector fields on a compact manifold M. A vector field f associates to each $x \in M$ an element of the tangent space of M at x, T_xM, and it does so continuously in x. The index of isolated zeros of f can again be defined. The following is very useful; it is known as the Poincaré–Hopf theorem and was proved in a general form by Hopf (1926*b*).

Theorem 2.4.2 *Suppose that f is a vector field on the compact oriented manifold M and suppose that f has isolated zeros, all of which are interior points of M. The sum of the indices of f at its zeros depends only on M (not on f); it is the Euler characteristic $\chi(M)$ $\left(=\sum_{0}^{m}(-1)^i \operatorname{rank} H_i(M)\right.$, where $H_i(M)$ is the* ith *homology group of M).*

We can deduce from Theorem 2.4.2 that, if n is even, then every vector field on the sphere S^n must have at least one zero.

3
Some topological applications

In this chapter the theory so far developed is used to prove results which are in the main of a topological nature. Some of these are of considerable usefulness – in the study of differential equations, for example; others, such as the Jordan separation theorem, are proved more for their intrinsic interest, particularly as the proofs usually given depend on the techniques of algebraic topology.

3.1 The Brouwer fixed point theorem

Perhaps the most famous of all fixed point theorems is that first stated by Brouwer (1912); it is very widely used. We give two versions of the theorem: the first is the most usual enunciation, while the second is of somewhat wider application.

Theorem 3.1.1 *Let D be an open subset of $\mathbf{R}^n$ such that $\bar{D}$ is homeomorphic to the closed unit ball $\bar{B}$. If $\phi \in C(\bar{D})$ and $\phi(\bar{D}) \subset \bar{D}$, then ϕ has a fixed point in $\bar{D}$.*

Proof Let $h: \bar{D} \to \bar{B}$ be the homeomorphism. Let $\psi = h \circ \phi \circ h^{-1}$; ψ maps $\bar{B}$ into itself and is continuous. Now if $\psi(y) = y$ with $y \in \bar{B}$, there is $x \in \bar{D}$ such that $y = h(x)$, and $h \circ \phi(x) = h(x)$; since h is a homeomorphism and $\phi(x) \in \bar{D}$, it follows that $\phi(x) = x$. To prove the theorem it is consequently necessary only to show that ψ has a fixed point in $\bar{B}$.

If it happens that $\psi(x_0) = x_0$ for some $x_0 \in \partial B$ there is nothing further to prove. We therefore suppose that $\psi(x) \neq x$ for $x \in \partial B$. Consider the homotopy

$$h_t(x) = x - t\psi(x) \quad (x \in \bar{B},\ 0 \leqslant t \leqslant 1).$$

It is clear that if $x \in \partial B$ and $0 \leqslant t < 1$, then $t\psi(x) \in B$; hence

$$h_t(x) \neq 0 \quad (x \in \partial B,\ 0 \leqslant t < 1).$$

Since, by hypothesis, $0 \notin h_1(\partial B)$ we may apply Theorem 2.1.2 to the homotopy h_t. Writing I for the identity mapping, this gives

$$d(I-\psi, B, 0) = d(I, B, 0).$$

But $d(I, B, 0) = 1$ by Theorem 1.1.4; so by Theorem 2.1.1 there is $x \in B$ such that $\psi(x) = x$. As we have already indicated, $h^{-1}(x)$ is then a fixed point of the mapping ϕ in D.

The result just proved is frequently applied when D is a convex set. We recall the following definition.

Definition 3.1.2 The set S is *convex* if $\lambda x + (1-\lambda)y \in S$ whenever $x \in S$, $y \in S$ and $0 \leqslant \lambda \leqslant 1$.

Brouwer's theorem can be applied to convex sets because of the following fact.

Theorem 3.1.3 *Let $K \subset \mathbf{R}^n$ be a closed, bounded, convex set with non-empty interior. There is a homeomorphism h of $\mathbf{R}^n$ onto itself such that $h(K) = \bar{B}$, the closed unit ball.*

Proof Take $x_0 \in \operatorname{int} K$; if $x \neq x_0$ the line segment $[x_0, x]$ or its continuation beyond x meets the boundary of K in a uniquely determined point: $f(x)$, say. Define

$$h(x_0) = 0$$

$$h(x) = \frac{x - x_0}{|f(x) - x_0|} \quad (x \neq x_0).$$

We leave the reader to check that h maps K onto $\bar{B}$ and is a homeomorphism of $\mathbf{R}^n$ onto itself.

A convex subset K of $\mathbf{R}^n$ may, of course, not have interior points; this is effectively because K is embedded in Euclidean space of dimension higher than is appropriate. Let the *dimension* of K be the largest integer r such that K contains $r+1$ points $x_0, x_1, \ldots, x_r$ for which $\{x_1 - x_0, \ldots, x_r - x_0\}$ is a linearly independent set of vectors (in $\mathbf{R}^n$). The linear space spanned by $x_1 - x_0, \ldots, x_r - x_0$

is denoted by $\mathscr{L}(K)$; it may be checked that $\mathscr{L}(K)$ is uniquely determined by K. It also can be shown that the interior of K relative to the affine space $x_0 + \mathscr{L}(K)$ is indeed non-empty: this is the 'natural' ambient space of K. For further details concerning these remarks the reader is referred to Eggleston (1958).

In the next result the essential parts of the proof of Theorem 3.1.1 are to some extent isolated.

Theorem 3.1.4 *Let D be a bounded, open subset of $\mathbf{R}^n$ and $\phi \in C(\bar{D})$. If there is $w \in D$ such that for all $x \in \partial D$ and all $\mu > 1$*

$$\phi(x) - w \neq \mu(x - w), \tag{3.1.1}$$

then ϕ has a fixed point in $\bar{D}$.

Proof As before we suppose that ϕ has no fixed point in ∂D. Let

$$h_t(x) = x - w - t(\phi(x) - w) \quad (x \in \bar{D}, 0 \leqslant t \leqslant 1).$$

If $h_t(x) = 0$ with $x \in \partial D$ and $0 < t < 1$, then, for $t^{-1} = \mu > 1$,

$$\phi(x) - w = \mu(x - w),$$

contradicting (3.1.1); also $w \in D$, so that $0 \notin h_0(\partial D)$. So, by Theorem 2.1.2,

$$d(I - \phi, D, 0) = d(I - w, D, 0).$$

Since $w \in D$, $d(I - w, D, 0) = 1$; hence, by Theorem 2.1.1, there must be $\xi \in D$ such that $(I - \phi)(\xi) = 0$ – that is, $\phi(\xi) = \xi$.

Remark It should be noted that in this theorem it is the image of the boundary of D that is significant. The geometric interpretation of the condition (3.1.1) is that for some 'origin' w inside D, for no $x \in \partial D$ does $\phi(x)$ lie on the continuation of the line $[w, x]$ beyond x.

The next result is often described as the property of 'invariance of a normal'.

Theorem 3.1.5 *Let D be a bounded, open subset of $\mathbf{R}^n$ containing* *the origin; suppose that n is odd. If $\phi \in C(\bar{D})$ and $0 \notin \phi(\partial D)$, then there are $y \in \partial D$, $\lambda \neq 0$ such that $\phi(y) = \lambda y$.*

Proof Define homotopies h_t and k_t:

$$\left.\begin{aligned} h_t(x) &= (1-t)\phi(x) + tx \\ k_t(x) &= (1-t)\phi(x) - tx \end{aligned}\right\} \quad (x \in \bar{D},\ 0 \leqslant t \leqslant 1).$$

If no $y \in \partial D$ and $\lambda \neq 0$ can be found to satisfy $\phi(y) = \lambda y$, then $h_t(x) \neq 0$ and $k_t(x) \neq 0$ for $x \in \partial D$ and $0 < t \leqslant 1$. Since $0 \notin \phi(\partial D)$, $h_0(x)$ and $k_0(x)$ are also non-zero for $x \in \partial D$. Theorem 2.1.2 applied to h_t and k_t, respectively, gives

$$d(\phi, D, 0) = d(I, D, 0),$$

$$d(\phi, D, 0) = d(-I, D, 0).$$

Now $d(I, D, 0) = 1$, and it is easily seen from the definition of degree that $d(-I, D, 0) = (-1)^n$. We thus have $1 = (-1)^n$, whence n is even, contrary to our hypothesis.

The condition that the dimension of the space is odd is necessary for the validity of the last theorem. A counter-example is afforded by the mapping of the unit disc in $\mathbf{R}^2$ given in polar co-ordinates by

$$(r, \theta) \mapsto (r, \theta + r).$$

Illustration We give a straightforward example of the application of Theorem 3.1.5 to the study of ordinary differential equations (see Kotin (1968)). Consider the equation

$$\dot{x} = f(x, t) \quad (x \in \mathbf{R}^n), \tag{3.1.2}$$

where f is continuous, homogeneous in x and periodic (of period ω, say) in t; that is, $f(\lambda x, t) = \lambda f(x, t)$ for $\lambda \neq 0$ and $f(x, t + \omega) = f(x, t)$. Suppose that for all x_0, t_0 there is one and only one solution of (3.1.2) satisfying $x = x_0$ when $t = t_0$; we then say that f is *admissible*. Let $x(t; t_0, x_0)$ be the solution such that $x(t_0; t_0, x_0) = x_0$. Since $x = 0$ is a constant solution of (3.1.2), the mapping

$$T: c \mapsto x(\omega; 0, c)$$

is defined on the ball $B_\alpha = B(0, \alpha)$ if α is sufficiently small; moreover, because f is admissible, $T(c) \neq 0$ if $c \neq 0$, and by a standard theorem on differential equations (see Coddington and Levinson (1955), Chapter 1, for example) T is continuous. (We shall have more to say about such equations as (3.1.2) in Chapter 9.)

If we now suppose that n is odd, Theorem 3.1.5 tells us that there are $y \in \partial B_\alpha$ and $\lambda \neq 0$ such that

$$x(\omega;0,y) = \lambda y. \tag{3.1.3}$$

Now $x(\omega + t;0,y)$ and $\lambda x(t;0,y)$ are both solutions of (3.1.2) (because of the homogeneity of f), and they coincide at $t = 0$ (by (3.1.3)). Therefore

$$x(\omega + t;0,y) = \lambda x(t;0,y)$$

for all t within the interval of definition of $x(t;0,y)$. We have thus shown the existence of a non-trivial solution $\phi(t)$ satisfying $\phi(t+\omega) = \lambda\phi(t)$; such a solution is called a *Floquet solution.* The case of even values of n can be covered by extending the given system by adding the equation $\dot{x}_{n+1} = x_{n+1}$, and then applying the above result.

3.2 Odd mappings

This section is concerned with a family of results related to the so-called Borsuk odd mapping theorem; roughly this states that the degree of an odd mapping relative to a symmetric open set is itself odd – and so in particular non-zero. We say that a set D is *symmetric* if $x \in D$ implies $-x \in D$; a function ϕ defined on such a set is *odd* if $\phi(-x) = -\phi(x)$. We first need some results on the extension of continuous functions: ψ *extends* ϕ if $\psi = \phi$ on the domain of definition of ϕ, while the domain of definition of ψ properly contains that of ϕ. The Tietze extension theorem (Tietze, 1915) is used; we state the theorem for metric spaces – the result holds for normal topological spaces (see, for example, Kelley (1955)).

Theorem 3.2.1 *Let X be a metric space, A a closed subset of X, and f a continuous, bounded, real-valued function defined on A. Then there is a continuous function $g: X \to \mathbf{R}$ which coincides with f on A and is such that*

$$\sup_{x \in X} g(x) = \sup_{y \in A} f(y), \quad \inf_{x \in X} g(x) = \inf_{y \in A} f(y).$$

Remark By applying the result to each co-ordinate, Theorem 3.2.1 ensures that a continuous, bounded function $f: A \to \mathbf{R}^m$ can be extended to a continuous function defined on the whole of X.

Following the scheme of Schwartz (1969), we prove some lemmas leading to a result on the extension of continuous functions which is useful when dealing with odd functions. The first is straightforward, but of independent interest.

Lemma 3.2.2 *Let K be a compact subset of $\mathbf{R}^n$, and $\phi \in C^1(K, \mathbf{R}^m)$ with $m > n$. Then $\phi(K)$ has measure zero in $\mathbf{R}^m$.*

Proof Since ϕ is continuously differentiable and K is compact, ϕ satisfies a Lipschitz condition:

$$|\phi(x) - \phi(y)| \leqslant \lambda |x - y| \quad (x, y \in K).$$

This is easily verified by considering the mapping ψ of the compact set $K \times K$ into $\mathbf{R}^m$ defined by

$$\psi(x, y) = \begin{cases} \dfrac{\phi(x) - \phi(y)}{|x - y|} & (x \neq y). \\ \phi'(x) & (x = y). \end{cases}$$

ψ is continuous and therefore bounded for $x, y \in K$. We regard $\mathbf{R}^n$ as a subset of $\mathbf{R}^m$. Now K has measure zero as a subset of $\mathbf{R}^m$; so given $\varepsilon > 0$, there are open balls $B_k (k = 1, 2, \ldots)$ in $\mathbf{R}^m$, of radii r_k, such that $K \subset \bigcup_k B_k$ and

$$\sum_k \text{measure}(B_k) < \varepsilon/\lambda^m.$$

Using the Lipschitz condition, it is seen that $\phi(B_k)$ lies in a ball C_k of radius λr_k; hence $\phi(K) \subset \bigcup_k C_k$ and

$$\sum_k \text{measure}(C_k) < \varepsilon.$$

Thus $\phi(K)$ has measure zero.

Lemma 3.2.3 *Let K and M be compact subsets of $\mathbf{R}^n$ with $K \subset M$; suppose that $m > n$ and $\phi: K \to \mathbf{R}^m$ is continuous and nowhere zero. Then ϕ can be extended to a mapping $\chi: M \to \mathbf{R}^m$ which is also continuous and nowhere zero.*

Proof Let $c = \inf\{|\phi(x)|; x \in K\}$, and take $\varepsilon < \frac{1}{2}c$. By Theorem 3.2.1 and the subsequent remark, ϕ can be extended to a continuous function ϕ_1 on $\mathbf{R}^n$; choose $\phi_2 \in C^1(M, \mathbf{R}^m)$ such that

$$|\phi_2(x) - \phi_1(x)| < \tfrac{1}{2}\varepsilon \quad (x \in M).$$

By Lemma 3.2.2, $\phi_2(M)$ has measure zero; so choose $p \in \mathbf{R}^m$ that $|p| < \frac{1}{2}\varepsilon$ and $p \notin \phi_2(M)$. Now let ψ be given by

$$\psi(x) = \phi_2(x) - p \quad (x \in M).$$

Then ψ is continuously differentiable, ψ is non-zero on M and

$$|\psi(x) - \phi(x)| < \varepsilon \quad (x \in K).$$

Choose $\eta: \mathbf{R}^1 \to \mathbf{R}^1$ to be continuous and to satisfy

$$\eta(t) = \begin{cases} 1 & (t \geqslant \frac{1}{2}c) \\ 2t/c & (t < \frac{1}{2}c). \end{cases}$$

Then define

$$\Phi(x) = \frac{\psi(x)}{\eta(|\psi(x)|)} \quad (x \in M).$$

It is clear that $|\Phi(x)| \geqslant \frac{1}{2}c$ $(x \in M)$ and, since $|\psi(x)| \geqslant \frac{1}{2}c$ $(x \in K)$, $\Phi = \psi$ on K, whence

$$|\Phi(x) - \phi(x)| < \varepsilon \quad (x \in K).$$

Applying Theorem 3.2.1 again, there exists a continuous function $\alpha: M \to \mathbf{R}^m$ such that

$$\alpha(x) = \Phi(x) - \phi(x) \quad (x \in K)$$

and $|\alpha(x)| \leqslant \varepsilon$ $(x \in M)$. Finally let $\chi = \Phi - \alpha$; then

$$\chi(x) = \phi(x) \qquad (x \in K)$$

and

$$|\chi(x)| \geqslant \tfrac{1}{2}c - \varepsilon > 0 \quad (x \in M).$$

Before continuing with the next lemma we establish the usage that 0 denotes the origin of $\mathbf{R}^n$.

Lemma 3.2.4 *Let $D \subset \mathbf{R}^n$ be an open, bounded, symmetric set with $0 \notin \bar{D}$. Suppose that $m > n$ and that $\phi: \partial D \to \mathbf{R}^m$ is a continuous mapping which is odd and nowhere zero. Then ϕ has an extension $\psi: \bar{D} \to \mathbf{R}^m$ which is also continuous, odd and nowhere zero.*

Proof The proof is by induction on the dimension n. First suppose that $n = 1$. So choose ε and N that $D \subset [-N, -\varepsilon] \cup [\varepsilon, N]$. Let

$$\phi_1 = \phi|_{\partial D \cap [\varepsilon, N]};$$

by Lemma 3.2.3, ϕ_1 can be extended to a continuous function ϕ_2 defined on $[\varepsilon, N]$ which is nowhere zero. Define $\psi: \bar{D} \to \mathbf{R}^m$ by

$$\psi(x) = \begin{cases} \phi_2(x) & (x \in \bar{D} \cap [\varepsilon, N]) \\ -\phi_2(-x) & (x \in \bar{D} \cap [-N, -\varepsilon]). \end{cases}$$

Then ψ is continuous, odd and nowhere zero.

Now let us suppose that the result holds for $n \leqslant k$. A function $\phi: \partial D \to \mathbf{R}^m$ is supposed given, where $D \subset \mathbf{R}^{k+1}$ satisfies the stated hypotheses, and $m > k + 1$; it is sought to extend ϕ to $\bar{D}$. It is convenient to identify $\mathbf{R}^k$ with $\mathbf{R}^{k+1} \cap \{x_{k+1} = 0\}$. Let ϕ_1 be the restriction of ϕ to $\partial D \cap \mathbf{R}^k$; by the induction hypothesis ϕ_1 can be extended to an odd, nowhere vanishing, continuous function $\phi_2: \bar{D} \cap \mathbf{R}^k \to \mathbf{R}^m$. Define $\phi_3: (\bar{D} \cap \mathbf{R}^k) \cup \partial D \to \mathbf{R}^m$ to coincide with ϕ_2 on $\bar{D} \cap \mathbf{R}^k$ and with ϕ on ∂D in such a way that it is continuous, odd and nowhere zero. Let

$$\mathbf{R}^{k+1}_{\pm} = \mathbf{R}^{k+1} \cap \{x_{k+1} \gtrless 0\}$$

$$D_{\pm} = \mathbf{R}^{k+1}_{\pm} \cap D,$$

and

$$\hat{D} = (\bar{D} \cap \mathbf{R}^k) \cup \partial D \cup D_+ .$$

By Lemma 3.2.3, ϕ_3 can be extended to a continuous function ϕ_4 from the compact set $\hat{D} \subset \mathbf{R}^{k+1}$ into $\mathbf{R}^m$ which is nowhere zero. Finally define $\psi: \bar{D} \to \mathbf{R}^m$ by

$$\psi(x) = \begin{cases} \phi_4(x) & (x \in \hat{D}) \\ -\phi_4(-x) & (x \in \bar{D} \backslash \hat{D}). \end{cases}$$

Then ψ is an odd function which is nowhere zero and coincides with ϕ on ∂D; it is clear that ψ is also continuous. The result is thus proved.

Theorem 3.2.5 *Let $D \subset \mathbf{R}^n$ be bounded, open and symmetric, with $0 \notin \bar{D}$. Let $\phi: \partial D \to \mathbf{R}^n$ be continuous, odd and nowhere zero. Then ϕ can be extended to a function $\psi: \bar{D} \to \mathbf{R}^n$ which is continuous and odd on $\bar{D}$, and non-zero on $\bar{D} \cap \mathbf{R}^{n-1}$.*

Note Our convention in this theorem and in the proof of the next is that $\mathbf{R}^{n-1}$ is identified with $\mathbf{R}^n \cap \{x_n = 0\}$. $\mathbf{R}^n_+$ and $\mathbf{R}^n_-$ will be used to denote $\mathbf{R}^n \cap \{x_n > 0\}$ and $\mathbf{R}^n \cap \{x_n < 0\}$, respectively.

Proof Let ϕ_1 be the restriction of ϕ to $\partial D \cap \mathbf{R}^{n-1}$; by Lemma 3.2.4, ϕ_1 has an extension ϕ_2 to $\bar{D} \cap \mathbf{R}^{n-1}$ which is continuous, odd and nowhere zero. Now define $\phi_3 : (\bar{D} \cap \mathbf{R}^{n-1}) \cup \partial D \to \mathbf{R}^n$ by

$$\phi_3(x) = \begin{cases} \phi_2(x) & (x \in \bar{D} \cap \mathbf{R}^{n-1}) \\ \phi(x) & (x \in \partial D). \end{cases}$$

Clearly ϕ_3 is continuous, odd and nowhere zero. Tietze's theorem (Theorem 3.2.1) enables us to extend ϕ_3 to a continuous function $\phi_4 : (\bar{D} \cap \mathbf{R}^n_+) \cup \partial D \to \mathbf{R}^n$. Finally the required function ψ is defined by

$$\psi(x) = \begin{cases} \phi_4(x) & (x \in \bar{D} \cap \mathbf{R}^n_+) \\ -\phi_4(-x) & (x \in \bar{D} \backslash (\bar{D} \cap \mathbf{R}^n_+)). \end{cases}$$

Again ψ is continuous and odd; it is non-zero on $\bar{D} \cap \mathbf{R}^{n-1}$.

We are now in a position to state and prove the main result of this section, namely the odd mapping theorem.

Theorem 3.2.6 *Let D be a bounded, open, symmetric subset of $\mathbf{R}^n$ containing the origin 0. If $\phi: \bar{D} \to \mathbf{R}^n$ is continuous, $0 \notin \phi(\partial D)$, and for all $x \in \partial D$*

$$\frac{\phi(x)}{|\phi(x)|} \neq \frac{\phi(-x)}{|\phi(-x)|}, \tag{3.2.1}$$

then $d(\phi, D, 0)$ is an odd number.

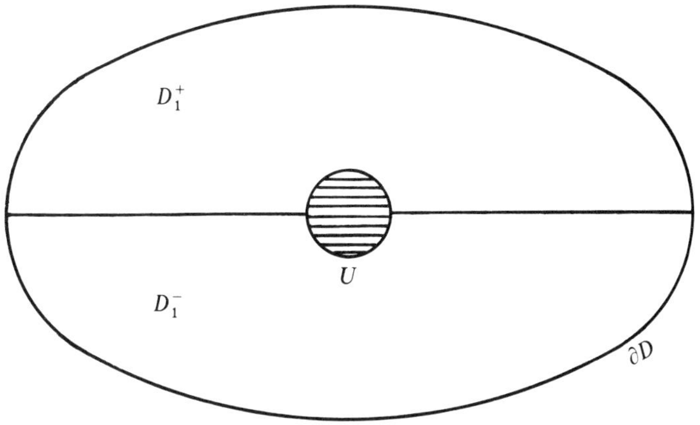

Figure 3.1. The odd mapping theorem

Remark The geometric meaning of the condition (3.2.1) is that, for $x \in \partial D$, the vectors $\phi(x)$ and $\phi(-x)$ are never in the same direction. Note that if ϕ is odd, (3.2.1) follows from the requirement that $0 \notin \phi(\partial D)$.

Proof The proof is usefully subdivided into five sections; the reader may find that Figure 3.1 clarifies matters somewhat.

(*a*) Let $\qquad \psi(x) = \phi(x) - \phi(-x) \quad (x \in \bar{D})$

and $\qquad H(t, x) = \phi(x) - t\phi(-x) \quad (x \in \bar{D},\ 0 \leqslant t \leqslant 1).$

Then, if $H(t, x) = 0$ with $x \in \partial D$, we have $\phi(x) = t\phi(-x)$; this contradicts the condition (3.2.1). So (by Theorem 2.1.2)

$$d(\phi, D, 0) = d(\psi, D, 0).$$

Thus it is necessary to prove the result of Theorem 3.2.6 only for functions which are odd. Henceforth we assume ϕ itself is odd.

(*b*) Let $U = \bar{B}(0, \varepsilon) \subset D$. Define $\phi_1 : U \cup \partial D \to \mathbf{R}^n$ by

$$\phi_1(x) = \begin{cases} \phi(x) & (x \in \partial D) \\ x & (x \in U). \end{cases}$$

We write $D_1 = D \backslash U$. If ϕ_2 denotes the restriction of ϕ_1 to ∂D_1, we may apply Theorem 3.2.5 to obtain a continuous function

$\phi_3 : \bar{D}_1 \to \mathbf{R}^n$ which is an extension of ϕ_2, is odd, and is non-zero on $\bar{D}_1 \cap \mathbf{R}^{n-1}$. Now set

$$\phi_4(x) = \begin{cases} \phi_3(x) & (x \in \bar{D}_1) \\ x & (x \in U); \end{cases}$$

ϕ_4 is continuous. Since $0 \notin \phi(\partial D)$ and $\phi_4 = \phi_3 = \phi_2 = \phi_1 = \phi$ on ∂D, we have

$$d(\phi_4, D, 0) = d(\phi, D, 0). \tag{3.2.2}$$

(*c*) Write $D_1^+ = \mathbf{R}^n_+ \cap D_1$ and $D_1^- = \mathbf{R}^n_- \cap D_1$. Since $\phi_3(x) \neq 0$ when $x \in \bar{D}_1 \cap \mathbf{R}^{n-1}$, we have, by the excision property (Theorem 2.2.1(2)) and 'domain decomposition',

$$d(\phi_3, D_1, 0) = d(\phi_3, D_1^+, 0) + d(\phi_3, D_1^-, 0). \tag{3.2.3}$$

We now exploit the fact that degree is invariant under a non-singular change of co-ordinates (Theorem 1.4.3). By making the transformation $(x_1, \ldots, x_n) \mapsto (-x_1, \ldots, -x_n)$, D_1^+ and D_1^- are interchanged, and because ϕ_3 is odd, $\phi_3|_{D_1^-}$ is simply $\phi_3|_{D_1^+}$ expressed in the new co-ordinates. Therefore

$$d(\phi_3, D_1^+, 0) = d(\phi_3, D_1^-, 0),$$

whence $d(\phi_3, D_1, 0)$ is even (using (3.2.3)).

(*d*) Referring back to (*b*), and in particular to the choice of ϕ_4, we have, writing U^0 for the interior of U,

$$\begin{aligned} d(\phi, D, 0) &= d(\phi_4, D, 0) \quad ((3.2.2)) \\ &= d(\phi_4, U^0 \cup (D \setminus U), 0) \quad (\text{Theorem } 2.2.1(2)) \\ &= d(\phi_4, U^0, 0) + d(\phi_4, D_1, 0) \quad (\text{Theorem } 2.2.1(1)) \\ &= 1 + d(\phi_4, D_1, 0) \quad (\phi_4|_U = I) \\ &= 1 + d(\phi_3, D_1, 0) \quad (\phi_4|_{\bar{D}_1} = \phi_3). \end{aligned} \tag{3.2.4}$$

(*e*) Using (3.2.4) and the conclusion of (*c*) that $d(\phi_3, D_1, 0)$ is even, we obtain the required result that $d(\phi, D, 0)$ is odd.

We proceed to use Theorem 3.2.6 to prove a slight generalisation of Borsuk's fixed point theorem (Borsuk, 1933).

Theorem 3.2.7 *Let D be a bounded, open, symmetric subset of $\mathbf{R}^n$ containing 0. Let $\phi:\partial D \to \mathbf{R}^m$ be continuous, and $m < n$. Then there is $x \in \partial D$ such that $\phi(x) = \phi(-x)$.*

Proof We identify $\mathbf{R}^m$ with $\mathbf{R}^n \cap \{(x_1, \dots, x_n); x_n = \dots = x_{m+1} = 0\}$. Define

$$\psi(x) = \phi(x) - \phi(-x) \quad (x \in \partial D).$$

Then ψ is an odd function and $\psi(\partial D) \subset \mathbf{R}^m$. Let $\psi = (\psi_1, \dots, \psi_m)$, and let $\hat{\psi}_1, \dots, \hat{\psi}_m$ be extensions of $\psi_1, \dots, \psi_m$ to $\bar{D}$. Define $\chi: \bar{D} \to \mathbf{R}^n$ by $\chi = (\hat{\psi}_1, \dots, \hat{\psi}_m, 0, \dots, 0)$. The function χ is continuous, and is odd on ∂D. Suppose that $0 \notin \psi(\partial D)$; then $0 \notin \chi(\partial D)$ and

$$\frac{\chi(x)}{|\chi(x)|} \neq \frac{\chi(-x)}{|\chi(-x)|} \quad (x \in \partial D).$$

Hence, by Theorem 3.2.6, $d(\chi, D, 0) \neq 0$. For $\varepsilon \neq 0$, write $p_\varepsilon = (0, \dots, 0, \varepsilon) \in \mathbf{R}^n \setminus \mathbf{R}^m$; if ε is sufficiently small,

$$d(\chi, D, p_\varepsilon) = d(\chi, D, 0) \neq 0.$$

Therefore $p_\varepsilon \in \chi(D)$. But $\chi(D) \subset \mathbf{R}^m$; we have a contradiction, so we conclude that $0 \in \psi(\partial D)$. The result is thus proved.

We give two corollaries of Theorem 3.2.7. The first is known as the Borsuk–Ulam theorem (and requires no further proof); we shall have recourse to the second in Chapter 6.

Corollary 3.2.8 *Let $S^n \subset \mathbf{R}^{n+1}$ be the n-sphere $x_1^2 + \dots + x_{n+1}^2 = 1$, and let $\phi: S^n \to \mathbf{R}^n$ be continuous. There is $\xi \in S^n$ with $\phi(\xi) = \phi(-\xi)$.*

Corollary 3.2.9 *Suppose that S^n is covered by n subsets $A_1, \dots, A_n$, all closed in S^n. Some A_i must contain a pair of antipodal points (that is, there is a set A_i and there is $\xi \in A_i$ such that $-\xi \in A_i$ also).*

Proof For $x \in S^n$, let $f_i(x)$ be the distance from x to A_i (calculated in $\mathbf{R}^{n+1}$); let

$$f(x) = (f_1(x), \dots, f_n(x)).$$

Then $f: S^n \to \mathbf{R}^n$ is continuous. By Theorem 3.2.7 or Corollary 3.2.8, there is $\xi \in S^n$ such that $f(\xi) = f(-\xi)$. Now ξ is contained in one of the $A_i : A_k$, say; since in that case $f_k(\xi) = 0$, it follows that $f_k(-\xi) = 0$. But A_k is closed, so that $-\xi \in A_k$.

Finally in this section we use Borsuk's fixed point theorem to prove the result known as the 'ham sandwich theorem'. Although the terms 'measurable' and 'hyperplane' have not been explicitly defined in this book, we shall use them without explanation in this theorem – they do not appear in the sequel, and this is not the context in which to investigate them more deeply.

Theorem 3.2.10 *Let $X_1, \ldots, X_n$ be bounded, measurable subsets of $\mathbf{R}^n$. There exists a hyperplane Y in $\mathbf{R}^n$ of dimension $n-1$ which bisects each of $X_1, \ldots, X_n$.*

Proof For $x \in S^n$, let π_x be the n-dimensional hyperplane in $\mathbf{R}^{n+1}$ through $\alpha = (0, \ldots, 0, 1)$ and orthogonal to the line segment $[0, x]$. For $1 \leqslant r \leqslant n$ let $f_r(x)$ be the measure of that part of X_r which lies on the same side of π_x as $(x + \alpha)$ (see Figure 3.2). Now π_x changes continuously with x; so $f_r : S^n \to \mathbf{R}$ is continuous ($r = 1, \ldots, n$). Thus

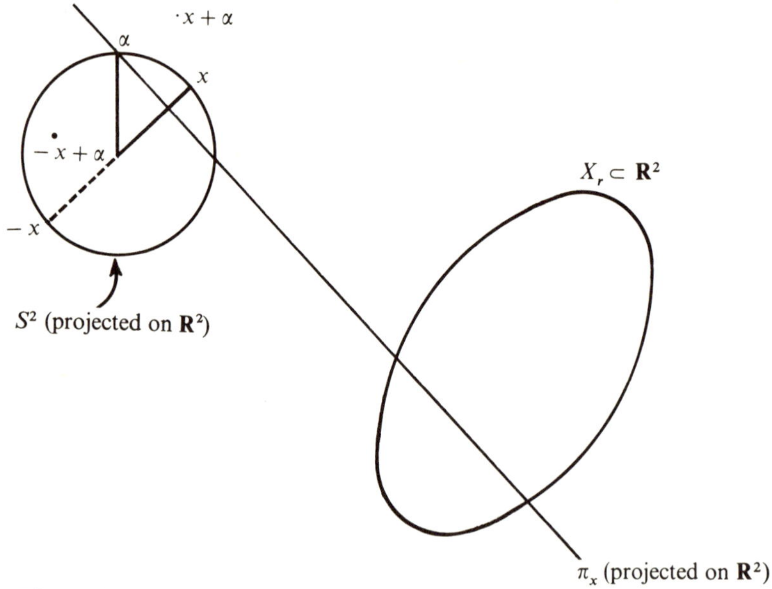

Figure 3.2. The ham sandwich theorem

$$f=(f_1,\ldots,f_n):S^n\to\mathbf{R}^n$$

is also continuous. By Theorem 3.2.7, there is $y\in S^n$ such that $f(y)=f(-y)$. Since, for all $x, \pi_x=\pi_{-x}$ and $x+\alpha$, $-x+\alpha$ are on opposite sides of π_x, it follows that π_y bisects each X_r. The required hyperplane in $\mathbf{R}^n$ is simply $\pi_y\cap\mathbf{R}^n$.

3.3 The Jordan separation theorem

In this section degree theory is used to prove the n-dimensional analogue of the Jordan curve theorem. There are, of course, other methods of proof; the degree theoretic proof utilises the multiplication theorem (Theorem 2.3.1), and is due to Leray (1950).

Theorem 3.3.1 *Let K and L be compact subsets of $\mathbf{R}^n$ which are homeomorphic. Then either $\mathscr{C}K$ and $\mathscr{C}L$ have the same finite number of components, or both have countably infinitely many.*

Proof It is helpful to break the proof up into a sequence of steps; Figure 3.3 illustrates the argument.

(*a*) Let $h:K\to L$ be the homeomorphism whose existence is given; let ϕ,ψ be continuous functions defined on the whole of $\mathbf{R}^n$ which extend h, h^{-1}, respectively (the existence of such extensions is assured by, say, Tietze's theorem). Both the sets $\mathscr{C}K$ and $\mathscr{C}L$ are open, so they have countably many components: this follows from the fact that a point with rational co-ordinates can be chosen in each component. Let the components of $\mathscr{C}K$ be Δ_s $(s=0,1,2,\ldots)$ and let those of $\mathscr{C}L$ be $D_r (r=0,1,2,\ldots)$; since K and L are compact,

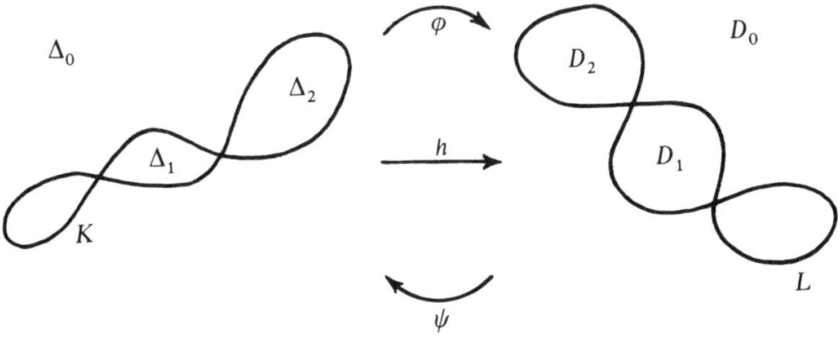

Figure 3.3. The Jordan separation theorem

one and only one of the Δ_s is unbounded, and one and only one of the D_r – we take the unbounded components to be Δ_0 and D_0. Now choose

$$p \in \mathbf{R}^n \backslash \psi \circ \phi(\partial \Delta_s);$$

if $s \neq 0$, we have

$$d(\psi \circ \phi, \Delta_s, p) = d(I, \Delta_s, p), \tag{3.3.1}$$

for $\psi \circ \phi = I$ on $\partial \Delta_s$. (As usual, I denotes the identity map.) If $p \in \Delta_t$ then $p \notin \partial \Delta_s$ for all s; for $s \neq 0$, we have

$$d(I, \Delta_s, \Delta_t) = \delta_{st}, \tag{3.3.2}$$

where $d(I, \Delta_s, \Delta_t)$ is, according to the convention established in Section 2.1, the value of $d(I, \Delta_s, p)$ when $p \in \Delta_t$.

(*b*) We shall use Theorem 2.3.1 to prove that, for all i and $j > 0$,

$$\sum_{k=1}^{\infty} d(\psi, D_k, \Delta_i) d(\phi, \Delta_j, D_k) = \delta_{ij} \tag{3.3.3}$$

and

$$\sum_{k=1}^{\infty} d(\phi, \Delta_k, D_i) d(\psi, D_j, \Delta_k) = \delta_{ij}. \tag{3.3.4}$$

The proof of these equations will occupy us in (*c*), (*d*) and (*e*).

(*c*) Since $\mathscr{C}\phi(\partial \Delta_j)$ $(j = 0, 1, 2, \ldots)$ are open sets, they all have countably many components: let $G_l^{(j)}$ $(l = 0, 1, 2, \ldots)$ be the components of $\mathscr{C}\phi(\partial \Delta_j)$, and let the unbounded component be $G_0^{(j)}$. Thus

$$\mathscr{C}L = \bigcup_{r=0}^{\infty} D_r, \quad \mathscr{C}\phi(\partial \Delta_j) = \bigcup_{l=0}^{\infty} G_l^{(j)} \quad (j = 0, 1, 2, \ldots).$$

Now, for every j, $\mathscr{C}L \subset \mathscr{C}\phi(\partial \Delta_j)$; so each of the D_r is a subset, possibly proper, of one of the $G_l^{(j)}$ $(l = 0, 1, 2, \ldots)$ – and $D_0 \subset G_0^{(j)}$. We shall now take $j \geqslant 1$ and work with a fixed value of j. The family $\{D_r; r \geqslant 1\}$ may be relabelled $\{D_{lk}^{(j)}; l \geqslant 0, k \geqslant 1\}$ in such a way that

$$U_0^{(j)} \equiv D_0 \cup D_{01}^{(j)} \cup D_{02}^{(j)} \cup \ldots \cup D_{0t}^{(j)} \cup \ldots \subset G_0^{(j)},$$
$$\cdot \quad \cdot \quad \cdot$$
$$U_l^{(j)} \equiv D_{l1}^{(j)} \cup D_{l2}^{(j)} \cup \ldots \cup D_{lt}^{(j)} \cup \ldots \subset G_l^{(j)},$$
$$\cdot \quad \cdot \quad \cdot$$

The relabelling depends in general on the value of j; also the $D_{lk}^{(j)}$

are disjoint (with j, of course, fixed). We therefore have, for all $k \geqslant 1$ and $l \geqslant 0$,

$$d(\phi, \Delta_j, D_{lk}^{(j)}) = d(\phi, \Delta_j, G_l^{(j)}). \tag{3.3.5}$$

(*d*) Retaining our fixed value of j, let

$$M_i^{(j)} = \overline{G_i^{(j)}} \backslash U_i^{(j)} \quad (i = 0, 1, \ldots).$$

If $x \in M_i^{(j)}$, then $x \notin D_0$ and $x \notin D_{lk}^{(j)}$ for all $l \geqslant 0$ and $k \geqslant 1$; hence $x \notin \bigcup_{r=0}^{\infty} D_r = \mathscr{C} L$. Thus $M_i^{(j)} \subset L$ $(i = 0, 1, \ldots)$. Now take $p \in \Delta_i (i > 0)$; $p \notin K = \psi(L)$. By the excision property (Theorem 2.2.1(2)), therefore,

$$d(\psi, G_l^{(j)}, p) = d(\psi, U_l^{(j)}, p) \quad (l = 1, 2, \ldots).$$

It follows that

$$\begin{aligned} d(\psi, G_l^{(j)}, p) &= d(\psi, \bigcup_{k=1}^{\infty} D_{lk}^{(j)}, p) \\ &= \sum_{k=1}^{\infty} d(\psi, D_{lk}^{(j)}, p), \end{aligned} \tag{3.3.6}$$

the last step by the property of domain decomposition (Theorem 2.2.1(1)).

(*e*) Using the multiplication theorem (Theorem 2.3.1), we obtain (still with $p \in \Delta_i$)

$$d(\psi \circ \phi, \Delta_j, p) = \sum_{l=1}^{\infty} d(\psi, G_l^{(j)}, p) d(\phi, \Delta_j, G_l^{(j)}); \tag{3.3.7}$$

this is a finite summation. The three equations (3.3.5), (3.3.6) and (3.3.7) together yield

$$d(\psi \circ \phi, \Delta_j, p) = \sum_{k=1}^{\infty} \sum_{l=1}^{\infty} d(\psi, D_{lk}^{(j)}, p) d(\phi, \Delta_j, D_{lk}^{(j)}). \tag{3.3.8}$$

But $\{D_{lk}^{(j)}; k \geqslant 1, l \geqslant 0\} = \{D_r; r = 1, 2, \ldots\}$, and $d(\phi, \Delta_j, D_{lk}^{(j)}) = 0$ if $l = 0$. Thus

$$d(\psi \circ \phi, \Delta_j, p) = \sum_{k=1}^{\infty} d(\psi, D_k, p) d(\phi, \Delta_j, D_k).$$

Now (3.3.1) and (3.3.2) give

$$d(\psi\circ\phi, \Delta_j, \Delta_i) = \delta_{ij}.$$

Since p is an arbitrary point of Δ_i, we have

$$\sum_{k=1}^{\infty} d(\psi, D_k, \Delta_i)d(\phi, \Delta_j, D_k) = \delta_{ij},$$

which is (3.3.3).

(f) By reversing the rôles of the Δs and the Ds the same argument as above leads to (3.3.4).

(g) Suppose that both the families $\{\Delta_s\}$ and $\{D_r\}$ are finite. Define matrices $A = (\alpha_{rs})$ and $B = (\beta_{rs})$:

$$\alpha_{rs} = d(\psi, D_r, \Delta_s), \quad \beta_{rs} = d(\phi, \Delta_r, D_s) \quad (r, s \geqslant 1).$$

If there are R of the D_r and S of the Δ_s, (3.3.3) and (3.3.4) can be written

$$AB = I_R, \quad BA = I_S,$$

where I_R, I_S are, respectively, the identity $R\times R$ and $S\times S$ matrices. We have $R = \text{rank}(AB) \leqslant \text{rank}(A) \leqslant S$, and $S = \text{rank}(BA) \leqslant \text{rank}(B) \leqslant R$; thus $R = S$.

(h) A similar argument to that of (g) shows that it is not possible to have finitely many D_r and infinitely many Δ_s nor finitely many Δ_s and infinitely many D_r.

The proof of Theorem 3.3.1 is now complete.

One of the consequences of Theorem 3.3.1 is the result known as 'invariance of domain'. A proof using the odd mapping theorem is possible; in the present context, however, a proof such as the following is simpler.

Theorem 3.3.2 *Let D be an open subset of $\mathbf{R}^n$ (not necessarily bounded). If $\phi: D\to\mathbf{R}^n$ is one to one and continuous, then $\phi(D)$ is open.*

Proof Take $p\in D$ and let B be an open ball with centre p such that $\bar{B}\subset D$. Since $\bar{B}$ is compact and ϕ is one to one and continuous, $\phi|_{\bar{B}}$ and $\phi|_{\partial B}$ are both homeomorphisms onto their images. By

Theorem 3.3.1, $\mathscr{C}\phi(\bar{B})$ is connected and $\mathscr{C}\phi(\partial B)$ has two components – Δ_1 and Δ_2, say, where Δ_1 is bounded and Δ_2 is unbounded. Since ϕ is one to one, $\phi(B)$ is a subset of $\mathscr{C}\phi(\partial B)$, and it is connected; therefore $\phi(B) \subset \Delta_1$ or $\phi(B) \subset \Delta_2$. If there is $q \in \Delta_1 \setminus \phi(\bar{B})$, there are paths in $\mathscr{C}\phi(\bar{B})$ joining q to the points of $\mathscr{C}\phi(\bar{B}) \cap \Delta_2$; such paths intersect $\phi(\partial B)$ – a contradiction. Therefore $\Delta_1 \subset \phi(\bar{B})$. From this we deduce that $\phi(B) \subset \Delta_1$. Hence $\phi(p) \in \Delta_1 \subset \phi(\bar{B}) \subset \phi(D)$; thus $\phi(D)$ is an open set.

Remark We can view Theorem 3.3.2 as saying that if $E \subset \mathbf{R}^n$, $\phi \in C(\bar{E})$ and ϕ is a homeomorphism of E onto its image $\phi(E)$, then ϕ maps the interior of E onto the interior of $\phi(E)$, and the boundary of E onto the boundary of $\phi(E)$. This is of relevance to the use of the Brouwer fixed point theorem in the formulation of Theorem 3.1.1.

We can use Theorems 3.3.1 and 3.3.2 to calculate the degree of injective mappings.

Theorem 3.3.3 *Let D be an open, bounded subset of $\mathbf{R}^n$ and let $\phi \in C(\bar{D})$ be one to one. If $p \in \phi(D)$, then $d(\phi, D, p) = \pm 1$.*

Proof It follows from Theorem 3.3.2 that ϕ is a homeomorphism of D onto $\phi(D)$. Take a ball B in $\phi(D)$ with centre p; obviously $\phi^{-1}(B)$ is connected. We apply the multiplication theorem (Theorem 2.3.1) to the composition $\phi \circ \phi^{-1} : B \to \phi(D)$. Remembering the remark following Theorem 3.3.2, we have

$$d(\phi \circ \phi^{-1}, B, p) = d(\phi^{-1}, B, \Delta) d(\phi, \Delta, p), \tag{3.3.9}$$

where Δ is the bounded component of $\mathscr{C}\phi^{-1}(\partial B)$. It is clear that $d(\phi \circ \phi^{-1}, B, p) = 1$; $d(\phi, \Delta, p) = d(\phi, D, p)$ by the excision property and the injectivity of ϕ. So (3.3.9) becomes

$$1 = d(\phi, D, p) d(\phi^{-1}, B, \Delta).$$

This is possible only if $d(\phi, D, p) = \pm 1$.

4
Leray-Schauder degree

4.1 Introductory remarks

We now wish to extend the theory of the previous chapters to spaces which are infinite dimensional. In some ways the most natural setting is a locally convex topological vector space; for the sake of simplicity of exposition, however, we shall present the theory in the context of a normed linear space. We shall see later that there is little loss of generality in so doing. The methods used can with little difficulty be adapted to the requirements of locally convex spaces.

Let us suppose, then, that $(X, \|\,.\,\|)$ is a normed linear space. Our notation will be analogous to that of the earlier chapters; in particular D will denote an open, bounded subset of X with closure $\bar{D}$ and boundary ∂D, and p will be a point of X. We seek to define an integer $d(\phi, D, p)$ for a suitable class of functions $\phi : \bar{D} \to X$. Any reasonable definition of degree should have the following three properties:

(1) $d(I, D, p) = +1$ $(p \in D)$,

(2) $d(\phi, D, p) \neq 0$ implies $\phi(x) = p$ for some $x \in D$,

(3) if $h_t(x)$ is a homotopy with $p \notin h_t(\partial D)$ for $0 \leqslant t \leqslant 1$, then $d(h_t, D, p)$ is independent of t.

In finite dimensional spaces, degree was defined for the class of continuous mappings; in infinite dimensional spaces, however, some restriction has to be placed on the mapping. The following example, given by Leray (1936) and to be found in Cronin (1964), demonstrates this.

Example Let X be the Banach space of continuous functions x: $[0, 1] \to \mathbf{R}$ with the norm

$$\|x\| = \max_{0 \leqslant s \leqslant 1} |x(s)|.$$

Let x_0 be the constant function

$$x_0(s) = \tfrac{1}{2} \quad (0 \leqslant s \leqslant 1),$$

and let

$$D = \{x; x \in X, \| x - x_0 \| < \tfrac{1}{2}\}.$$

Clearly D is open and bounded. Choose $\phi \in X$ such that $\phi(0) = 0$, $\phi(1) = 1$ and $0 \leqslant \phi(s) \leqslant 1$ for $0 \leqslant s \leqslant 1$. Now define $\Phi : \bar{D} \to X$ by

$$\Phi(x) = \phi \circ x \quad (x \in \bar{D}).$$

Then $\Phi(\bar{D}) \subset \bar{D}$. We shall consider the homotopy

$$h_t(x) = t\Phi(x) + (1 - t)x \quad (x \in \bar{D}, 0 \leqslant t \leqslant 1).$$

If $y \in \partial D$, then $\| x_0 - y \| = \frac{1}{2}$, so that $0 \leqslant y(s) \leqslant 1$ for $s \in [0, 1]$, and for some $s_0 \in [0, 1]$ either $y(s_0) = 0$ or $y(s_0) = 1$. In the case $y(s_0) = 0$ we have $h_t(y)(s_0) = 0$ and in the case $y(s_0) = 1$ we have $h_t(y)(s_0) = 1$. Since $0 \leqslant \phi(s) \leqslant 1, 0 \leqslant h_t(y)(s) \leqslant 1$ also; hence $y \in \partial D$ implies that $h_t(y) \in \partial D$ for $t \in [0, 1]$.

Suppose now that the three desirable properties of a degree mentioned above hold. Take $Y \in D$. Since $h_t(\partial D) \subset \partial D$ for $t \in [0, 1]$, condition (3) applied to the homotopy h_t implies that

$$d(\Phi, D, Y) = d(I, D, Y).$$

Condition (1) tells us that $d(I, D, Y) = 1$. Finally condition (2) allows us to deduce that there is a solution in D of the equation

$$\Phi(x) = Y.$$

This conclusion, however, is erroneous. To see this, define the functions Y and ϕ as follows (see Figure 4.1):

$$Y(s) = \tfrac{1}{4} + \tfrac{1}{2}s$$

$$\phi(s) = \begin{cases} s & (0 \leqslant s \leqslant \frac{1}{2}) \\ 1 - s & (\frac{1}{2} < s \leqslant \frac{5}{8}) \\ \frac{5}{3}(s - 1) + 1 & (\frac{5}{8} < s \leqslant 1). \end{cases}$$

If $x(s)$ is a solution of $\Phi(x) = Y$, we have $\phi(x(0)) = \frac{1}{4}$, whence $x(0) =$

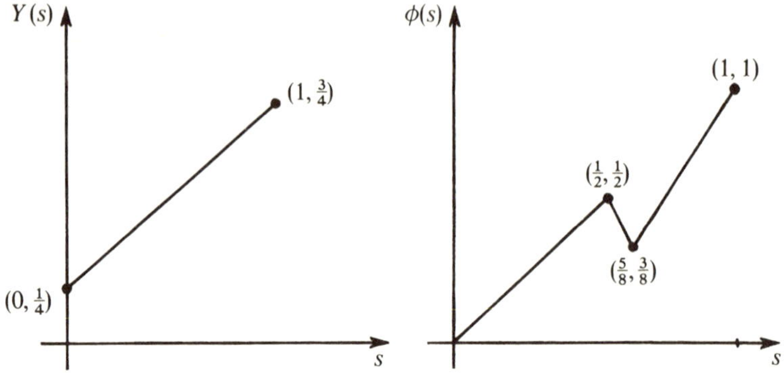

Figure 4.1. The graphs of Y and ϕ

$\frac{1}{4}$. Thus $\phi(x(s))$ can increase from $\frac{1}{4}$ by at most $\frac{1}{4}$ before it starts to decrease. But $Y(s)$ increases from $\frac{1}{4}$ to $\frac{3}{4}$. Therefore no suitable function $x(s)$ can be found in D.

It is not possible, therefore, to satisfactorily define a concept of degree for the class $C(\bar{D})$ of all continuous functions from $\bar{D}$ into X. In the next section we define degree for a particular subclass of $C(\bar{D})$.

4.2 Definition of the Leray–Schauder degree

We shall consider mappings of the form $I - T$, where I is the identity mapping and T is compact (see below). The degree we define is the Leray–Schauder degree, so described following its introduction by Leray and Schauder (1934). The technique is to approximate T by mappings of finite rank.

Definition 4.2.1 Let E and F be (real) normed spaces, and $M \subset E$. The mapping $T: M \to F$ is *compact* if (i) T is continuous, and (ii) for every bounded subset A of M, $T(A)$ is a relatively compact subset of F – that is, $\overline{T(A)}$ is compact.

Theorem 4.2.2 *Suppose that E and F are real normed spaces, M is a bounded subset of E, and $T: M \to F$ is compact. Given $\varepsilon > 0$, there is a continuous mapping $T_\varepsilon : M \to F$ whose range $T_\varepsilon(M)$ is finite dimensional such that*

$$\| T(u) - T_\varepsilon(u) \| < \varepsilon \quad (u \in M).$$

Proof. The set $\overline{T(M)}$ is compact, and so is covered by a finite number of balls in F of radius ε; let these be $B(v_i, \varepsilon)(i = 1, \ldots, N)$, where the $v_i \in \overline{T(M)}$. For $i = 1, \ldots, N$ and $x \in M$, define

$$m_i(x; \varepsilon) = \max\{0, \varepsilon - \| T(x) - v_i \| \}.$$

Each function m_i is continuous in x, and for a given x some m_i is non-zero. We can therefore define

$$\theta_i(x; \varepsilon) = \frac{m_i(x; \varepsilon)}{\sum_{k=1}^{N} m_k(x; \varepsilon)} \quad (i = 1, \ldots, N, x \in M).$$

Each function θ_i is continuous, and

$$\sum_{i=1}^{N} \theta_i(x; \varepsilon) = 1$$

for all x; in addition the support of θ_i is given by

$$\operatorname{supp} \theta_i = \overline{T^{-1}(B(v_i, \varepsilon))}.$$

Finally we define

$$T_\varepsilon(x) = \sum_{i=1}^{N} \theta_i(x; \varepsilon) v_i \quad (x \in M).$$

Then $T_\varepsilon : M \to F$ is continuous, and $T_\varepsilon(M)$ is contained in the finite dimensional linear space spanned by the vectors $v_1, \ldots, v_N$. Moreover

$$\| T(x) - T_\varepsilon(x) \| = \| \sum_{i=1}^{N} \theta_i(x, \varepsilon)(T(x) - v_i) \| \quad (x \in M).$$

Since $\theta_i = 0$ unless $\| T(x) - v_i \| < \varepsilon$, and $\Sigma \theta_i = 1$, it follows that

$$\| T(x) - T_\varepsilon(x) \| < \varepsilon \quad (x \in M).$$

We now return to the situation described at the beginning of §4.1: X is a normed linear space, D is an open, bounded subset of X, and $p \in X$. We consider $\phi = I - T$, where $T : \bar{D} \to X$ is compact. Having found a mapping T_ε which approximates T and which has finite dimensional range, we shall define $d(\phi, D, p)$ in terms of the degree of $I - T_\varepsilon$ relative to an appropriate finite dimensional subset of D. As remarked at the end of Chapter 1, the theory developed in the previous chapters applies to any finite dimensional normed space, and we saw in Theorem 1.4.3 that degree is invariant under a change of co-ordinates.

Before continuing, we need the following lemma. For $m < n$, $\mathbf{R}^m$ is identified with $\mathbf{R}^n \cap \{x \in \mathbf{R}^n; x_{m+1} = \ldots = x_n = 0\}$. In this lemma, ρ denotes the distance in $\mathbf{R}^n$.

Lemma 4.2.3 *Suppose that $m \leqslant n$, D is an open, bounded subset of $\mathbf{R}^n$, and $\phi \in C(\bar{D}, \mathbf{R}^m)$. Let $\psi: \bar{D} \to \mathbf{R}^n$ be defined by*

$$\psi(x) = x + \phi(x) \quad (x \in \bar{D}).$$

Write $D^m = \mathbf{R}^m \cap D$ and χ for the restriction of ψ to $\mathbf{R}^m \cap \bar{D}$. If $p \in \mathbf{R}^m \backslash \psi(\partial D)$, then

$$d(\psi, D, p) = d(\chi, D^m, p). \tag{4.2.1}$$

Proof The result is certainly true if $D^m = \emptyset$, for then $p \notin \psi(\bar{D})$, and both sides of (4.2.1) are zero. We suppose that $D^m \neq \emptyset$; χ maps $\overline{D^m}$ into $\mathbf{R}^m$. The right hand side of (4.2.1) is defined, for $\partial(\mathbf{R}^m \cap D) \subset \mathbf{R}^m \cap \partial D$, whence $p \notin \chi(\partial D^m)$. If $\psi(x) = p$, then $x = p - \phi(x) \in \mathbf{R}^m$; thus $\psi^{-1}(p) \subset D^m$, so that $\psi^{-1}(p) = \chi^{-1}(p)$.

Consider first the case when ϕ is a C^1 function and $p \notin$ crease χ. Then

$$\begin{aligned} d(\psi, D, p) &= \sum_{x \in \psi^{-1}(p)} \operatorname{sign} J_\psi(x) \\ &= \sum_{x \in \chi^{-1}(p)} \operatorname{sign} K(x), \end{aligned}$$

where $K(x)$ is the determinant of the partitioned matrix

$$\left[\begin{array}{c|c} J_\chi(x) & 0_m \\ \hline & I_{n-m} \end{array}\right];$$

here the notation is self-explanatory. Hence

$$d(\psi, D, p) = \sum_{x \in \chi^{-1}(p)} \operatorname{sign} J_\chi(x)$$

$$= d(\chi, D^m, p).$$

Now consider the general case. For $j = m+1, \ldots, n$ let $\hat{\phi}_j = 0$, and for $j = 1, \ldots, m$ choose $\hat{\phi}_j \in C^1(\bar{D}, \mathbf{R})$ so that, if $\hat{\phi} = (\hat{\phi}_1, \hat{\phi}_2, \ldots, \hat{\phi}_n)$,

$$|\hat{\phi}(x) - \phi(x)| < \rho(p, \psi(\partial D)) \quad (x \in \bar{D}).$$

Writing $\hat{\psi}(x) = x + \hat{\phi}(x)$, we also have

$$|\hat{\psi}(x) - \psi(x)| < \rho(p, \psi(\partial D)) \quad (x \in \bar{D}). \tag{4.2.2}$$

Let $\hat{\chi}$ be the restriction of $\hat{\psi}$ to $\overline{D^m}$. Since $\hat{\psi}^{-1}(p) \subset D^m$ and the crease of $\hat{\chi}$ has m-measure zero, we can, by making a small translation on $\hat{\phi}$, ensure that in addition to (4.2.2) we have

$$p \notin \text{crease } \hat{\chi}.$$

Then $d(\psi, D, p) = d(\hat{\psi}, D, p)$, and the result follows from the earlier part of the proof.

To define the Leray–Schauder degree we proceed in a number of steps. Recall that we are dealing with mappings $\phi : \bar{D} \to X$ of the form $\phi = I - T$, where T is compact. From now on ρ denotes the distance induced by the norm on X. We shall always suppose that $p \notin \phi(\partial D)$.

(*a*) Let $r = \rho(p, \phi(\partial D))$ – that is

$$r = \inf\{\|p - \phi(x)\|; x \in \partial D\}.$$

Then $r > 0$; for if not, there is a sequence (x_n) in ∂D such that $\phi(x_n) \to p$ as $n \to \infty$. The set $\{T(x_n)\}$ is relatively compact, for $\{x_n\}$ is a bounded set and T is a compact mapping; there is therefore a convergent subsequence of $(T(x_n))$. Without loss of generality, we may suppose that the sequence $(T(x_n))$ itself converges: $T(x_n) \to y$, say. Then $y \in \overline{T(D)}$, and

$$x_n = T(x_n) + \phi(x_n) \to y + p \quad (n \to \infty).$$

Now $x_n \in \partial D$, a closed set, so $y + p \in \partial D$. But by continuity

$$y = \lim_{n \to \infty} T(x_n) = T(y + p).$$

This implies that $\phi(y+p)=p$, that is, $p\in\phi(\partial D)$ – a contradiction. We thus have that $r>0$.

(*b*) Take ε with $0<\varepsilon<r$. Using Theorem 4.2.2, there is a mapping $T_\varepsilon:\bar{D}\to X$ with finite dimensional range such that

$$\|T(x)-T_\varepsilon(x)\|<\varepsilon \quad (x\in\bar{D}).$$

Let $\mathscr{S}_\varepsilon$ be the finite dimensional normed space spanned by $T_\varepsilon(\bar{D})$ and p:

$$\mathscr{S}_\varepsilon=\mathrm{Sp}\{T_\varepsilon(\bar{D}),p\}.$$

Let
$$D_\varepsilon=D\cap\mathscr{S}_\varepsilon$$
and
$$\phi_\varepsilon(x)=x-T_\varepsilon(x) \quad (x\in\bar{D}).$$

Then D_ε is a bounded, open subset of $\mathscr{S}_\varepsilon$; also $\partial_\varepsilon D_\varepsilon\subset\partial D$, where $\partial_\varepsilon D_\varepsilon$ is the boundary of D_ε in $\mathscr{S}_\varepsilon$. We see that $\phi_\varepsilon(\bar{D}_\varepsilon)\subset\mathscr{S}_\varepsilon$, and for $x\in\partial D$,

$$\|x-T_\varepsilon(x)-p\|\geqslant\|x-T(x)-p\|-\|T(x)-T_\varepsilon(x)\|$$

$$>r-\varepsilon>0.$$

It follows that $d(\phi_\varepsilon,D_\varepsilon,p)$ is defined; strictly ϕ_ε refers, of course, to the restriction of ϕ_ε to D_ε, but we avoid this pedantry here and in the sequel. If $D_\varepsilon=\emptyset$, then $d(\phi_\varepsilon,D_\varepsilon,p)=0$.

(*c*) **Lemma 4.2.4** *For* $0<\varepsilon<r, d(\phi_\varepsilon,D_\varepsilon,p)$ *is independent of* ε.

Proof Take ε and η both in the interval $(0,r)$. Let $\mathscr{S}_\varepsilon,\mathscr{S}_\eta$ be as defined above, and let

$$\mathscr{S}_\mu=\mathrm{Sp}\{\mathscr{S}_\varepsilon,\mathscr{S}_\eta\}.$$

Write $D_\mu=D\cap\mathscr{S}_\mu$; Lemma 4.2.3 then gives

$$\left.\begin{aligned} d(\phi_\varepsilon,D_\varepsilon,p)&=d(\phi_\varepsilon,D_\mu,p)\\ d(\phi_\eta,D_\eta,p)&=d(\phi_\eta,D_\mu,p).\end{aligned}\right\} \tag{4.2.3}$$

Consider the homotopy

$$h_t(x)=t\phi_\varepsilon(x)+(1-t)\phi_\eta(x) \quad (x\in\bar{D}_\mu, 0\leqslant t\leqslant 1).$$

Then

$$\|h_t(x)-\phi(x)\| \leqslant t\|\phi_\varepsilon(x)-\phi(x)\| + (1-t)\|\phi_\eta(x)-\phi(x)\|$$
$$< t\varepsilon + (1-t)\eta < r. \qquad (4.2.4)$$

So for $x \in \partial D_\mu$ we have

$$\|h_t(x)-p\| \geqslant \|\phi(x)-p\| - \|\phi(x)-h_t(x)\| > 0.$$

Hence $d(\phi_\varepsilon, D_\mu, p) = d(\phi_\eta, D_\mu, p)$; the result follows from this and (4.2.3).

(*d*) Suppose that V is any finite dimensional space containing $\mathscr{S}_\varepsilon$, where $0<\varepsilon<r$. With $D \cap V = D_V$, Lemma 4.2.3 ensures that $d(\phi_\varepsilon, D_V, p) = d(\phi_\varepsilon, D_\varepsilon, p)$. We can now give the definition of $d(\phi, D, p)$; it is justified by the remarks already made.

Definition 4.2.5 Suppose D is an open, bounded subset of X and $\phi = I - T$, where $T: \bar{D} \to X$ is compact; suppose also that $p \in X \setminus \phi(\partial D)$. Introduce $\hat{\phi} = I - \hat{T}$, where $\hat{T}$ is a continuous mapping defined in $\bar{D}$ with finite dimensional range, so chosen that

$$\|T(x) - \hat{T}(x)\| < \rho(p, \phi(\partial D)) \quad (x \in \bar{D}).$$

Choose a finite dimensional linear space V to contain $\hat{T}(\bar{D})$ and p; let $D_V = D \cap V$. Then define

$$d(\phi, D, p) = d(\hat{\phi}, D_V, p).$$

Remarks (1) Some authors describe a mapping $T: X \to X$ as compact if its range $T(X)$ is relatively compact. In this way an apparent generalisation of the definition of degree to the case when D is unbounded is obtained.

(2) On examining Definition 4.2.5 it is seen that D need not be bounded – it is only necessary that D meets every finite dimensional subspace of X in a bounded subset. Thus 'bounded' may throughout be replaced by 'finitely bounded': D is *finitely bounded* if $D \cap V$ is bounded in V for all finite dimensional linear subspaces V of X.

(3) The definition of degree remains possible if T has the property of mapping bounded sets to precompact sets. This is generally

weaker than being compact; the two, however, coincide if the space X is complete.

(4) We shall not again make explicit reference to remarks (1), (2) and (3). It is desirable that they be borne in mind, to be deployed in applications should the need arise.

4.3 Properties of the Leray–Schauder degree

In the transition from finite to infinite dimensional spaces the class of functions for which the concept of degree is defined has contracted substantially; not unexpectedly, however, virtually all the properties survive the process.

Throughout this chapter, p, D, ϕ and T will be as in Section 4.2 (in particular $\phi = I - T$ and T is compact). If M is a subset of X, let $K(M)$ denote the set of compact mappings from M into X and let

$$K_1(M) = \{\phi; \phi = I - T, T \in K(M)\},$$

that is $K_1(M)$ is the set of compact perturbations of the identity on M.

Theorem 4.3.1 *If $p \in D$, then $d(I, D, p) = 1$; if $p \notin \bar{D}$, then $d(I, D, p) = 0$.*

Proof It is only necessary to take the space V in Definition 4.2.5 to contain p and to let $\hat{T} = 0$; the result then follows from Theorem 1.1.4.

Theorem 4.3.2 *Let $\phi \in K_1(\bar{D})$ and $d(\phi, D, p) \neq 0$; then there is $x \in D$ such that $\phi(x) = p$.*

Proof For all suitably small ε, we can choose a mapping T_ε to approximate T as in Theorem 4.2.2; in the notation of Definition 4.2.5,

$$d(\phi, D, p) = d(\hat{\phi}, D_V, p).$$

Here D_V is contained in a finite dimensional normed space, $\hat{\phi}$ is defined on $\bar{D}$, and

$$\|\phi(x) - \hat{\phi}(x)\| < \varepsilon \quad (x \in \bar{D}).$$

Since $d(\phi, D, p) \neq 0$, there is, for all sufficiently large $n, x_n \in D$ such that

$$x_n - T_{n-1}(x_n) = p.$$

The set $\{x_n\}$ is bounded, so $\{T(x_n)\}$ is relatively compact. As usual we suppose that the sequence $(T(x_n))$ is itself convergent: say $T(x_n) \to \xi$. Now

$$\| x_n - T(x_n) - p \| = \| T_{n-1}(x_n) - T(x_n) \| < \frac{1}{n};$$

so $x_n \to \xi + p$ as $n \to \infty$. Since $T(x_n) \to \xi$, we see that $\xi = T(\xi + p)$, and so $\phi(\xi + p) = p$.

Next we look at the property of invariance under homotopy. We need the following definition

Definition 4.3.3 Given a subset M of X, suppose h maps the interval $[0, 1]$ into $K(M)$. We say that h is a *homotopy of compact transformations* on M if, given $\varepsilon > 0$ and a bounded subset L of M, there is $\delta(\varepsilon, L) > 0$ such that

$$\| (h(t))(x) - (h(s))(x) \| < \varepsilon \quad (x \in L, |t - s| < \delta).$$

Theorem 4.3.4 (Invariance under homotopy) *Let D (as usual) be a bounded, open subset of X, and let $h(t)$ be a homotopy of compact transformations on $\bar{D}$ such that, if $\phi_t = I - h(t)$,*

$$p \notin \phi_t(\partial D) \quad (0 \leqslant t \leqslant 1).$$

Then $d(\phi_t, D, p)$ is independent of $t \in [0, 1]$.

Proof First we show that there is $r > 0$ such that

$$\| \phi_t(x) - p \| \geqslant r \quad (x \in \partial D, 0 \leqslant t \leqslant 1). \tag{4.3.1}$$

For if not, there are sequences $(x_n), (t_n)$ in ∂D, $[0, 1]$, respectively, such that $\| \eta_n \| < n^{-1}$, where

$$\eta_n = \phi_{t_n}(x_n) - p.$$

Clearly (t_n) has a convergent subsequence – say $t_n \to \tau$ as $n \to \infty$. Since $h(\tau)$ is compact, $(h(\tau)(x_n))$ also has a convergent subsequence, and again we suppose that the whole sequence converges:

$$h(\tau)(x_n) \to y.$$

Since $h(t)$ is a homotopy of compact transformations on $\bar{D}$ and $\{x_m\}$ is bounded,

$$\| h(\tau)(x_n) - h(t_n)(x_n) \| \to 0 \quad (n \to \infty).$$

Hence $\| y - h(t_n)(x_n) \| \leqslant \| y - h(\tau)(x_n) \| + \| h(\tau)(x_n) - h(t_n)(x_n) \|$

$$\to 0.$$

Therefore $h(t_n)(x_n) \to y$ as $n \to \infty$, and

$$x_n = \eta_n + h(t_n)x_n + p \to y + p.$$

But $x_n \in \partial D$; then $y + p \in \partial D$, and we have

$$\begin{aligned} \phi_\tau(y+p) &= y + p - \lim_{n\to\infty} h(t_n)(x_n) \\ &= p, \end{aligned}$$

which is a contradiction. The inequality (4.3.1) is therefore proved.

Now define a relation on $[0, 1]$ as follows:

$$s \sim t \text{ if } d(\phi_s, D, p) = d(\phi_t, D, p). \tag{4.3.2}$$

It is easily seen that (4.3.2) defines an equivalence relation; we show that the equivalence classes are open subsets of $[0, 1]$. From this it follows that there is only one class. Take $\tau \in [0, 1]$ and choose ε so that $0 < \varepsilon < \frac{1}{4}r$ (r being as defined in the first part of the proof). Corresponding to ε, take a vector space V and a mapping $h_\varepsilon(\tau)$ to approximate $h(\tau)$ as in Definition 4.2.5; in particular

$$\| h_\varepsilon(\tau)(x) - h(\tau)(x) \| < \tfrac{1}{4}r \quad (x \in \bar{D}). \tag{4.3.3}$$

According to Definition 4.3.3, there is δ such that $|t - \tau| < \delta$ implies that

$$\| h(\tau)(x) - h(t)(x) \| < \tfrac{1}{4}r \quad (x \in \bar{D}). \tag{4.3.4}$$

Putting (4.3.3) and (4.3.4) together, we have

$$\| h(t)(x) - h_\varepsilon(\tau)(x) \| < \tfrac{1}{2}r \quad (|t - \tau| < \delta, x \in \bar{D}).$$

This means that $h_\varepsilon(\tau)$ can be used as an approximant for $h(t)$ in Definition 4.2.5; that is

$$d(\phi_t, D, p) = d(I - h_\varepsilon(\tau), D_V, p), \tag{4.3.5}$$

where $D_V = D \cap V$, and V is a suitably chosen finite dimensional space. But the right hand side of (4.3.5) is just $d(\phi_\tau, D, p)$; thus $t \sim \tau$ if $|t-\tau| < \delta$. This means that the equivalence classes of the relation $\sim$ defined by (4.3.2) are open, so completing the proof of the theorem.

Theorem 4.3.5 *Suppose that $\phi \in K_1(\bar{D})$, $\psi \in K_1(\bar{D})$, and $\phi = \psi$ on ∂D. Then, if $p \notin \phi(\partial D)$,*

$$d(\phi, D, p) = d(\psi, D, p).$$

Proof Let $\phi = I - T_1$ and $\psi = I - T_2$. Consider the homotopy $\phi_t = I - h(t)$, where

$$h(t)(x) = (1-t)T_1(x) + tT_2(x) \quad (x \in \bar{D}, 0 \leqslant t \leqslant 1).$$

Clearly $h(t)$ is a homotopy of compact transformations on $\bar{D}$, and if $x \in \partial D$, then $\phi_t(x) = (1-t)\phi(x) + t\psi(x) = \phi(x)$. Since $p \notin \phi(\partial D)$, the result follows at once from Theorem 4.3.4.

Theorem 4.3.6 *Suppose that $\phi \in K_1(D)$, $p \notin \phi(\partial D)$ and $q \in X$; let*

$$\phi_1(x) = \phi(x) - q \quad (x \in \bar{D}).$$

Then

$$d(\phi, D, p) = d(\phi_1, D, p - q).$$

Proof Let $\phi = I - T$; then $\phi_1 = I - T_1$, where $T_1(x) = T(x) + q$. Clearly, $\phi_1 \in K_1(\bar{D})$. Suppose $\hat{T}$ approximates T as required in the definition of the Leray–Schauder degree: that is, $\hat{T}$ has finite dimensional range and

$$\|\hat{T}(x) - T(x)\| < r = \rho(p, \phi(\partial D)) \quad (x \in \bar{D}).$$

Let $\hat{T}_1(x) = \hat{T}(x) + q$; then

$$\|\hat{T}_1(x) - T_1(x)\| < r = \rho(p - q, \phi_1(\partial D)) \quad (x \in \bar{D}).$$

Take a finite dimensional space V to contain $\hat{T}(D)$, $\hat{T}_1(D), p$ and $p - q$; let $D_V = D \cap V$. Then, if $\hat{\phi} = I - \hat{T}$ and $\hat{\phi}_1 = I - \hat{T}_1$,

$$d(\phi, D, p) = d(\hat{\phi}, D_V, p) \quad \text{and} \quad d(\phi_1, D, p-q) = d(\hat{\phi}_1, D_V, p-q),$$

both by Definition 4.2.5. But by Theorem 2.1.6

$$d(\hat{\phi}_1, D_V, p-q) = d(\hat{\phi}, D_V, p).$$

The result follows.

Theorem 4.3.7 *Suppose $\phi \in K_1(\bar{D})$ and $p \notin \phi(\partial D)$. If $\psi \in K_1(\bar{D})$ and*

$$\| \phi(x) - \psi(x) \| < r = \rho(p, \phi(\partial D)) \ (x \in \bar{D}),$$

then $p \notin \psi(\partial D)$ and

$$d(\phi, D, p) = d(\psi, D, p).$$

Proof This result follows almost immediately from Theorem 4.3.4. For consider the homotopy of compact perturbations of the identity given by

$$h_t(x) = t\psi(x) + (1-t)\phi(x) \quad (x \in \bar{D}, 0 \leqslant t \leqslant 1).$$

If $x \in \partial D$,

$$\begin{aligned} \| t\psi(x) + (1-t)\phi(x) - p \| &= \| t(\psi(x) - \phi(x)) + (\phi(x) - p) \| \\ &\geqslant \| \phi(x) - p \| - t \| \psi(x) - \phi(x) \| \\ &> r(1-t). \end{aligned}$$

Thus $p \notin h_t(\partial D)$ $(0 \leqslant t \leqslant 1)$, and the result follows.

Theorem 4.3.8 *Suppose $\phi \in K_1(\bar{D})$; $d(\phi, D, p)$ is the same for all p in the same component of $\mathscr{C}\phi(\partial D)$.*

Proof Let the ball $B(p, \varepsilon)$ be contained in $\mathscr{C}\phi(\partial D)$; take $q \in B(p, \varepsilon)$. Then, by Theorem 4.3.6,

$$d(\phi, D, q) = d(\phi - (q-p), D, p).$$

By Theorem 4.3.7,

$$d(\phi - (q-p), D, p) = d(\phi, D, p)$$

if ε is sufficiently small. So the function $p \mapsto d(\phi, D, p)$ is continuous in $\mathscr{C}\phi(\partial D)$, and integer-valued; it is therefore constant on components of $\mathscr{C}\phi(\partial D)$.

The reader will by now have a sound idea of the way in which the theorems we proved about degree in finite dimensional spaces are adapted to apply to compact perturbations of the identity in normed linear spaces in general. We shall now list other properties of degree which hold good in the setting of this chapter; most of the proofs are omitted. The reader who wishes to construct proofs will find guidance in Schwartz (1969) and Nagumo (1951*b*).

Theorem 4.3.9 *Suppose that* $\phi \in K_1(\bar{D})$ *and* $p \notin \phi(\partial D)$.

(1) (Domain decomposition) *If D is the disjoint union of open sets $D_i (i = 1, 2, \ldots)$, then*

$$d(\phi, D, p) = \sum_i d(\phi, D_i, p).$$

(2) (Excision) *If $K \subset \bar{D}$ is closed and $p \notin \phi(K)$, then*

$$d(\phi, D, p) = d(\phi, D \backslash K, p).$$

Theorem 4.3.10 (Multiplication) *Let $\phi \in K_1(\bar{D})$ and M a bounded, open set containing $\phi(\bar{D})$. Let $\Delta = M \backslash \phi(\partial D)$ and let $\Delta_i (i = 1, 2, \ldots)$ be the components of Δ. If $\psi \in K_1(\bar{M})$ and $p \notin \psi(\phi(\partial D)) \cup \psi(\partial M)$, then*

$$d(\psi \circ \phi, D, p) = \sum_j d(\psi, \Delta_j, p) d(\phi, D, \Delta_j). \qquad (4.3.6)$$

Remarks Since ϕ is a compact perturbation of the identity, $\phi(\bar{D})$ is bounded, so that M can certainly be chosen as required in the theorem. That $\psi \circ \phi \in K_1(\bar{D})$ follows easily from the facts that $\phi \in K_1(\bar{D})$ and $\psi \in K_1(\bar{M})$. The notation $d(\phi, D, \Delta_j)$ has the same significance as in Chapter 2. The summation (4.3.6) is finite; Δ has countably many components because it is an open subset of a locally connected space.

For the next four theorems we shall suppose that X is complete.

Theorem 4.3.11 (Odd mappings) *Let D be an open, bounded, symmetric subset of the Banach space X, with $0 \in D$. If $\phi \in K_1(\bar{D})$ and*

$$\frac{\phi(x)}{\|\phi(x)\|} \neq \frac{\phi(-x)}{\|\phi(-x)\|} \quad (x \in \partial D),$$

then $d(\phi, D, 0)$ is odd.

In Chapter 3 we mentioned that the property of 'domain invariance' could be deduced from the odd mapping theorem. We shall prove the next theorem in order to display this connection.

Theorem 4.3.12 (Domain invariance) *Suppose that D is an open subset of the Banach space X and that $\phi \in K_1(\bar{D})$ is one to one. Then $\phi(D)$ is an open set.*

Proof Take $p \in \phi(D)$, and let B be a ball such that $\phi^{-1}(p) \in B \subset D$. By Theorem 4.3.8, there is $\varepsilon > 0$ such that $d(\phi, B, p) = d(\phi, B, q)$ if $\|q - p\| < \varepsilon$. If we can show that $d(\phi, B, p) \neq 0$, then it follows that there are q-points of ϕ in B for all q in a neighbourhood of p; thus $\phi(D) \supset \phi(B) \supset \{q; \|q - p\| < \varepsilon\}$, whence $\phi(D)$ is an open set. Without loss of generality, we take $p = \phi^{-1}(p) = 0$. To see that $d(\phi, B, 0) \neq 0$, we use the homotopy

$$h_t(x) = \phi\left(\frac{x}{1+t}\right) - \phi\left(\frac{-tx}{1+t}\right) \quad (x \in \bar{B}, 0 \leqslant t \leqslant 1).$$

Clearly $h_t \in K_1(\bar{B})\,(0 \leqslant t \leqslant 1)$. Moreover, since ϕ is one to one, $h_t(x) = 0$ implies that $x = 0$. By Theorem 4.3.4, $d(\phi, B, 0) = d(\psi, B, 0)$, where ψ is the odd function $\phi(\tfrac{1}{2}x) - \phi(-\tfrac{1}{2}x)$. By Theorem 4.3.11, $d(\psi, B, p)$ is odd, and so non-zero. The proof is complete.

Theorem 4.3.13 (Separation theorem) *Suppose that D and D_1 are bounded, open subsets of the Banach space X such that there is a homeomorphism $h: \bar{D} \to \bar{D}_1$ which belongs to $K_1(\bar{D})$. Then $\mathscr{C}\bar{D}$ and $\mathscr{C}\bar{D}_1$ have the same finite number of components, or both have countably infinitely many.*

Theorem 4.3.14 (Homeomorphism) *Suppose that D is a bounded, open subset of the Banach space X and that $\phi \in K_1(\bar{D})$ is a one to* *one mapping of $\bar{D}$ onto $\phi(\bar{D})$. If $p \in \phi(D)$, then $d(\phi, D, p) = \pm 1$.*

Finally we give the analogue of Theorem 2.2.4.

Theorem 4.3.15 *Suppose that D_* is a bounded, open subset of $[0, 1] \times X$, and that $f: D_* \to X$ is compact. Let ϕ_t denote the mapping $x \mapsto x - f(t, x)$, and let $D_t = \{x \in X; (t, x) \in D\}$. If $p \notin \phi_t(\partial D_t)$ for $0 \leqslant t \leqslant 1$, then $d(\phi_t, D_t, p)$ is independent of t in $[0, 1]$.*

4.4 Fixed point theorems

In this section we prove a number of fixed point theorems which correspond in form to Brouwer's theorem in finite dimensional spaces. For a wider selection of such results, we refer the reader to the book of Smart (1974). Formulations of Theorems 4.4.2, 4.4.3, 4.4.6 and 4.4.10 were given by Schauder (1930), Rothe (1937), and Tychonoff (1935); Schaefer (1955) proved a form of Theorem 4.4.11.

Definition 4.4.1 A subset S of X has the *fixed point property* if every continuous mapping of S into itself has a fixed point.

Theorem 4.4.2 *Let S be a bounded, closed, convex subset of X containing the origin 0 in its interior. Let $\phi \in K(S)$ be such that $\phi(S) \subset S$. Then ϕ has a fixed point in S.*

Proof Let $D = \text{int } S$; since S is convex and $D \neq \emptyset$, $\bar{D} = S$ and $\partial D = \partial S$. Consider the homotopy

$$h_t(x) = x - t\phi(x) \quad (x \in \bar{D}, 0 \leqslant t \leqslant 1).$$

We may suppose that $\phi(x) \neq x$ for $x \in \partial D$, for otherwise there would be nothing further to prove; thus $0 \notin h_1(\partial D)$. For $x \in \partial D$ and $0 \leqslant t < 1$, it is easily seen that $t\phi(x) \in \text{int } S = D$; it follows that $0 \notin h_t(\partial D)$ for $0 \leqslant t \leqslant 1$. Using Theorem 4.3.4 we obtain

$$d(I - \phi, D, 0) = d(I, D, 0).$$

By Theorem 4.3.1, $d(I, D, 0) = 1$; therefore, by Theorem 4.3.2, there is $x \in D$ with $\phi(x) = x$.

The next result is a more general form of Theorem 4.4.2 in that some of the hypotheses are relaxed. In it (*a*) S need not be convex, (*b*) the condition $0 \in \operatorname{int} S$ is replaced by the requirement that $\operatorname{int} S$ is non-empty, and (*c*) it is the image of the boundary ∂S under ϕ that alone is significant. We shall later be able to relax the hypothesis that S has non-empty interior.

Theorem 4.4.3 *Let S be a bounded, closed subset of X with non-empty interior; let $\phi \in K(S)$. If there is $w \in \operatorname{int} S = D$ such that for all $\lambda > 1$ and $x \in \partial D$*

$$\phi(x) - w \neq \lambda(x - w), \tag{4.4.1}$$

then ϕ has a fixed point.

Proof We consider the homotopy

$$h_t(x) = x - w - t(\phi(x) - w) \quad (x \in \bar{D}, 0 \leqslant t \leqslant 1).$$

As before, we suppose that ϕ has no fixed point in ∂D – that is, $0 \notin h_1(\partial D)$. If $h_t(x) = 0$, with $0 < t < 1$ and $x \in \partial D$, then (4.4.1) is contravened. Hence $0 \notin h_t(\partial D)$ for $0 \leqslant t \leqslant 1$. Therefore, by Theorem 4.3.4,

$$d(I - \phi, D, 0) = d(I - w, D, 0).$$

Since $w \in D, d(I - w, D, 0) = 1$; thus there is $x \in D$ with $\phi(x) = x$.

Remark Note that condition (4.4.1) relates to $\partial(\operatorname{int} S)$; it is, of course, true that $\partial(\operatorname{int} S) \subset \partial S$. The hypothesis, and so the conclusion, of the theorem holds if S is convex and $\phi(\partial S) \subset S$.

In the two theorems given above the set S was required to have a non-empty interior. This restriction we remove by using the extension of Tietze's theorem given by Dugundji (1951). We state a version of this result which suits our needs, and omit the proof.

Definition 4.4.4 The *convex hull* of a set S, written $\operatorname{co} S$, is the intersection of all convex sets containing S. The *closed convex hull* of S, $\overline{\operatorname{co}}\, S$, is the intersection of all closed, convex sets containing S.

Remark co S is the smallest convex set containing S, $\overline{\text{co}}\ S$ is the smallest closed, convex set containing S, and $\overline{\text{co}}\ S = \overline{\text{co}\ S}$.

Theorem 4.4.5 (Dugundji) *Suppose that S is a closed subset of a metric space X, and L is a normed linear space. Every continuous function $f: S \to L$ has a continuous extension $F: X \to L$ such that $F(X) \subset \text{co} f(S)$.*

Theorem 4.4.6 *Let S be a bounded, closed, non-empty, convex subset of the normed space X, and let $\phi: S \to S$ be compact. Then ϕ has a fixed point.*

Proof Since S is bounded, it is contained in a ball B with centre at the origin 0. By Theorem 4.4.5, there is a continuous function $r: \bar{B} \to S$ such that $r|_S = I$, the identity. Let $\hat{\phi} = \phi \circ r$; $\hat{\phi}$ maps $\bar{B}$ into itself and is compact (because ϕ is compact). Thus the hypotheses of Theorem 4.4.2 are satisfied by $\hat{\phi}$ and $\bar{B}$. There is therefore $\xi \in \bar{B}$ such that $\hat{\phi}(\xi) = \xi$. But $\hat{\phi}(\bar{B}) \subset S$, whence $\xi \in S$ and $\phi(\xi) = \xi$.

Corollary 4.4.7 *Let S be as in the theorem and h be a homeomorphism of S onto a bounded subset S_1 of X. If $\phi: S_1 \to S_1$ is compact, then ϕ has a fixed point in S_1.*

Proof Write $\psi = h^{-1} \circ \phi \circ h$; ψ maps S into itself, and is compact. There is, therefore, $x \in S$ such that $\psi(x) = x$; it follows that $y = h(x)$ is a fixed point of ϕ in S_1.

Corollary 4.4.8 *A compact, convex subset S of a normed space X has the fixed point property.*

Proof If $\phi: S \to S$ is continuous, then it is also compact (for S is compact). By Theorem 4.4.6, ϕ has a fixed point in S.

Remark In contrast to Corollary 4.4.8, Dugundji (1951) showed that the unit ball of a normed space has the fixed point property if and only if the ambient space X is finite dimensional – so that Brouwer's theorem is false in infinite dimensional spaces.

We give one more result related to those above. We first quote a result which may be found in, for example, Dunford and Schwartz (1958, p. 416).

Lemma 4.4.9 *If X is a Banach space and $S \subset X$ is compact, then $\overline{\text{co}}\, S$ is also compact.*

Theorem 4.4.10 *Suppose that X is a Banach space. Let S be a closed, convex subset of X (not necessarily bounded), and ϕ be a continuous mapping of S into a compact subset of S. Then ϕ has a fixed point.*

Proof There is a compact set A such that $\phi(S) \subset A \subset S$. Write $\overline{\text{co}}\, A = A_0$; by Lemma 4.4.9, A_0 is compact. Also, since $A_0 \subset S$, $\phi(A_0) \subset A_0$. The result follows from Corollary 4.4.8 applied to A_0.

Finally we prove a result from which Schaefer's theorem (1955) follows as a corollary. The reader may wish to explore the possible variants of the stated theorem, in the spirit of the previous parts of this section.

Theorem 4.4.11 *Suppose that X is a normed space (possibly incomplete). Let S be a bounded, closed, convex subset of X, containing the origin 0 in its interior D. Let $H:[0,1] \times S \to X$ be a homotopy of compact transformations such that $H(0, \partial S) \subset S$ and*

$$H(t,x) \neq x \quad (0 \leqslant t < 1,\ x \in \partial S).$$

Then $\phi = H(1,.)$ has a fixed point in S.

Proof Since S is convex and $D \neq \emptyset$, $\bar{D} = S$ and $\partial D = \partial S$. Consider the homotopy

$$h_t(x) = x - H(t,x) \quad (x \in \bar{D},\ 0 \leqslant t \leqslant 1).$$

Our hypothesis is that $0 \notin h_t(\partial D)$ for $0 \leqslant t < 1$. Either there is a fixed point of ϕ in ∂D or else $0 \notin h_t(\partial D)$ for $0 \leqslant t \leqslant 1$. In the former case, there is nothing to prove; we may therefore assume the latter. Then, by Theorem 4.3.4,

$$d(h_1, D, 0) = d(h_0, D, 0). \tag{4.4.2}$$

To calculate $d(h_0, D, 0)$, we argue as in Theorem 4.4.3 and use the homotopy

$$k_\tau(x) = x - \tau H(0,x) \quad (x \in \bar{D},\ 0 \leqslant \tau \leqslant 1).$$

We are supposing that $H(0, \partial S) \subset S$, and also that $H(0, x) \neq x$ for $x \in \partial S$; since S is convex, it follows that $0 \notin k_\tau(\partial D)$ for $0 \leqslant \tau \leqslant 1$. Another application of Theorem 4.3.4 yields

$$d(k_0, D, 0) = d(k_1, D, 0).$$

Now $d(k_1, D, 0) = d(h_0, D, 0)$ and $d(k_0, D, 0) = 1$ (since $0 \in D$). From this and (4.4.2) we deduce that $d(h_1, D, 0) = 1$, whence there is $x \in D$ such that $x = \phi(x)$. The proof is complete.

Corollary 4.4.12 (Schaefer) *Let $\phi: X \to X$ be compact, and suppose that the set*

$$S = \{u; u = \lambda\phi(u), \text{ some } \lambda \in [0, 1)\}$$

is bounded. Then ϕ has a fixed point.

Proof Let B be a ball with centre 0 and such that $S \subset B$. Define $H: [0, 1] \times \bar{B} \to X$ by

$$H(t, x) = t\phi(x).$$

H is a homotopy of compact transformations, $H(0, \partial B) \subset B$ (trivially) and, since $S \subset B$,

$$H(t, x) \neq x \quad (0 \leqslant t < 1, x \in \partial B).$$

By the theorem, $H(1, .) = \phi$ has a fixed point.

5
Axiomatic treatment

In the following chapters we shall give definitions of degree which are of greater generality than those encountered up to now. The reader will have noticed that the later properties of the Leray–Schauder degree were derived from the definition and the earlier properties in much the same way as they were in the case of the Brouwer degree (that is, in finite dimensional spaces). The same feature would be apparent should we treat each of our subsequent generalisations *ab initio*. It would therefore be economic of effort and also aesthetically desirable if an axiomatic scheme were devised; then, once a given definition of degree were known to satisfy a small number of axioms, all the other properties expected to hold for a degree would follow from the general theory.

The question of finding a convenient set of axioms is connected with the problem of uniqueness. By this we mean the following: if a degree d_1 is defined for a class $\mathscr{F}_1$ of functions, and another degree d_2 is defined for a class $\mathscr{F}_2$, is it the case that $d_1 = d_2$ for functions in $\mathscr{F}_1 \cap \mathscr{F}_2$? In other words, do our loosely termed 'generalisations' merit the name?

In this chapter we outline an axiomatic theory of degree, and consider the uniqueness of the Leray–Schauder degree. We follow the approach of Amann and Weiss (1973), but work only in normed linear spaces. We shall see that little generality is lost by developing degree theory in normed spaces rather than in a wider class of topological vector spaces.

5.1 Axioms for degree theory

Let X be a real normed vector space; let τ denote the norm topology and ρ the induced metric. The closure, interior, boundary and complement of $S \subset X$ are written $\bar{S}$, int S, ∂S, $\mathscr{C}S$ respectively. Let

ω be a distinguished subset of τ containing the empty set $\emptyset$, and such that $\omega \neq \{\emptyset\}$; write $\omega\backslash\{\emptyset\} = \omega'$. For $D \in \omega$, let $C(\bar{D})$ denote the linear space of continuous mappings from $\bar{D}$ into X, with the topology of uniform convergence. To each $D \in \omega$ let there be assigned a subset $M(D)$ of $C(\bar{D})$ satisfying the following conditions.

(1) $I|_{\bar{D}} \in M(D)$ for all $D \in \omega'$.

(2) If $D, D_1 \in \omega$ and $D \subset D_1$, then $\phi|_D \in M(D)$ whenever $\phi \in M(D_1)$.

(3) If $\phi \in M(D)$, then, for all $p \in X$, $\phi - p \in M(D)$ also.

(As usual, I is the identity mapping and $\phi - p$ denotes the mapping $x \mapsto \phi(x) - p; C(\emptyset) = \emptyset$.)

The collection

$$M(\omega) = \{M(D); D \in \omega\}$$

is said to be an *admissible* class of mappings on X.

Examples The most common choice is $\omega = \beta$, the collection of bounded, open subsets of X; other possibilities are $\omega = \tau$ and $\omega = \gamma$, the collection of finitely bounded, open subsets of X. As for $M(D)$, it will often be $C(\bar{D})$ or $K_1(\bar{D})$, the set of compact perturbations of the identity on $\bar{D}$.

Definition 5.1.1 Let ω and $M(\omega)$ be given. With each triplet (ϕ, D, p) such that $D \in \omega$, $\phi \in M(D)$ and $p \in X\backslash\overline{\phi(\partial D)}$, let there be associated an integer $d(\phi, D, p)$. The correspondence d is a *topological degree* for $M(\omega)$ if the following axioms are satisfied (in which case we shall describe $d(\phi, D, p)$ as the degree of ϕ at p relative to D):

(I) If $D \in \omega'$ and $p \in D$, then $d(I|_{\bar{D}}, D, p) = 1$.

(II) If $D \in \omega'$, D_1 and D_2 are disjoint subsets of D with $D_1, D_2 \in \omega$, and $\phi \in M(D)$ with $p \notin \overline{\phi(\bar{D}\backslash(D_1 \cup D_2))}$, then

$$d(\phi, D, p) = d(\phi|_{\bar{D}_1}, D_1, p) + d(\phi|_{\bar{D}_2}, D_2, p).$$

(III) If $D \in \omega'$ and $h: [0, 1] \to M(D)$ is continuous, then, provided

$$p \notin \overline{h(t)(\partial D)} \quad (0 \leqslant t \leqslant 1),$$

$d(h(t), D, p)$ is independent of $t \in [0, 1]$.

(IV) If $D \in \omega'$, $\phi \in M(D)$ and $p \notin \overline{\phi(\partial D)}$, then

$$d(\phi, D, p) = d(\phi - p, D, 0).$$

Notation If $\phi \in M(D)$, $D_1 \subset D$ and $p \notin \overline{\phi(\partial D_1)}$, it is convenient to write $d(\phi, D_1, p)$ for $d(\phi|_{\bar{D}_1}, D_1, p)$. The relation in (II), for example, becomes

$$d(\phi, D, p) = d(\phi, D_1, p) + d(\phi, D_2, p).$$

Remarks (1) Note that in (II), it follows from the hypothesis $p \notin \overline{\phi(\bar{D} \setminus (D_1 \cup D_2))}$ that $p \notin \overline{\phi(\partial D_1)} \cup \overline{\phi(\partial D_2)}$; similarly in (IV) $p \notin \overline{\phi(\partial D)}$ implies that $0 \notin \overline{(\phi - p)(\partial D)}$.

(2) The properties we have used as axioms have been shown to hold for both the Brouwer and Leray–Schauder degrees. In the case of the former, $X = \mathbf{R}^n$, $\omega = \beta$, $M(D) = C(\bar{D})$; for the latter X was infinite dimensional, $\omega = \beta$ and $M(D) = K_1(\bar{D})$.

(3) A topological degree does not always exist. For example, let X be an infinite dimensional normed space, $\omega = \beta$ and $M(D) = C(\bar{D})$; we saw at the beginning of Chapter 4 that a satisfactory definition is impossible. The second of the three desirable properties we required of a degree at that point will be deduced from our axioms; the other two are themselves axioms.

(4) Axiom (II) is a combination of the properties of domain decomposition and excision, and will sometimes be called the axiom of additivity; (III) is homotopy invariance (recall here that $C(\bar{D})$ has the topology of uniform convergence and that $M(D)$ has the induced topology).

(5) Axioms (II) and (III) can be deduced from the same statements with 0 (the origin) in place of p, together with (IV).

Theorem 5.1.2 *The axioms (III) and (IV) together are equivalent to the following:*

(III′) *If $D \in \omega'$ and $h: [0,1] \to M(D)$, $\theta: [0,1] \to X$ are both continuous, then, provided*

$$\theta(t) \notin \overline{h(t)(\partial D)} \quad (0 \leqslant t \leqslant 1),$$

$d(h(t), D, \theta(t))$ is independent of $t \in [0,1]$.

Proof Suppose that (III) and (IV) hold. Then

$$d(h(t), D, \theta(t)) = d(h(t) - \theta(t), D, 0) \quad (0 \leqslant t \leqslant 1).$$

Now $H: t \mapsto h(t) - \theta(t)$ is a continuous mapping of $[0,1]$ into $M(D)$, and $0 \notin \overline{H(t)(\partial D)}$; so $d(H(t), D, 0)$ is independent of $t \in [0, 1]$. Thus $d(h(t), D, \theta(t))$ is independent of t. Conversely, suppose (III′) holds. The special case $\theta(t) = p$ gives (III), and the choice $h(t) = \phi - tp, \theta(t) = (1 - t)p$ gives (IV).

It will sometimes be convenient to use (III′) instead of (III) and (IV); we shall still refer to (III′) as 'homotopy invariance'.

Having set up an axiom scheme, it is natural to enquire to what extent the axioms are independent. First define $d_1(\phi, D, p) = 0$ whenever $p \notin \overline{\phi(\partial D)}$. Clearly d_1 satisfies (II) and (III′), but not (I). If d is a degree on $M(\omega)$, define

$$d_2(\phi, D, p) = [d(\phi, D, p)]^3$$

(whenever $p \notin \overline{\phi(\partial D)}$); then d_2 satisfies (I) and (III′), but not (II). Thirdly define $d_3(\phi, D, p)$ to be 1 if $p \in D$ and there is U such that $p \in U \in \omega'$ and $\phi = I$ on U; define $d_3(\phi, D, p) = 0$ in all other cases. Then d_3 satisfies (I) and (II), but not (III′) – in fact, it satisfies neither (III) nor (IV). It is therefore seen that the axioms are independent. If it so happens that (I), (II) and (III) alone determine a unique degree on $M(\omega)$, then clearly (IV) is automatically satisfied. On the other hand, if d_1 and d_2 are two degrees which satisfy (I), (II) and (III), then (IV) need not be satisfied by every degree, as the following shows. Suppose that ψ and q are such that $d_1(\psi, D, q) \neq d_2(\psi, D, q)$; for all relevant ϕ and p, define

$$d(\phi, D, p) = \begin{cases} d_1(\phi, D, p) & (p \neq q) \\ d_2(\phi, D, q) & (p = q). \end{cases}$$

Now d satisfies (I), (II) and (III), but not (IV) – for

$$\begin{aligned} d(\psi, D, q) = d_2(\psi, D, q) \neq d_1(\psi, D, q) &= d_1(\psi - (q - p), D, p) \\ &= d(\psi - (q - p), D, p). \end{aligned}$$

We proceed to deduce the main properties of a topological degree from the axioms (compare with Chapter 2). These will be grouped together in convenient collections; some are almost trivial, but are included for completeness. We shall sometimes require $M(D)$ to be convex; in practice $M(D)$ is usually a linear subspace of $C(\bar{D})$.

We shall use ϕ, D, p 'generically': $p \in X$, $D \in \omega'$, $\phi \in M(D)$ and $p \notin \overline{\phi(\partial D)}$ (so that $D \neq \emptyset$ and $d(\phi, D, p)$ is defined).

Theorem 5.1.3 (1) *If there is $D \in \omega'$ such that $d(\phi, D, p)$ is defined, then $d(\phi, \emptyset, p) = 0$.*

(2) *Let K be a subset of $\bar{D}$ with $D \backslash K \in \omega$; if $p \notin \overline{\phi(K)}$, then*

$$d(\phi, D, p) = d(\phi, D \backslash K, p).$$

(3) *If D is the disjoint union of sets $D_1, \ldots, D_N$ in ω, then*

$$d(\phi, D, p) = \sum_{i=1}^{N} d(\phi, D_i, p).$$

Proof (1) Using axiom (II) with $D_1 = \emptyset$ and $D_2 = D$, we immediately have $d(\phi, \emptyset, p) = 0$.

(2) We again use (II), with $D_1 = D \backslash K$ and $D_2 = \emptyset$. Now D is an open set, so $\underline{\bar{D} \backslash (D \backslash K) = \partial D} \cup K$; since $p \notin \overline{\phi(K)}$ and $p \notin \overline{\phi(\partial D)}$, it follows that $p \notin \phi(\bar{D} \backslash (D \backslash K))$. Axiom (II) then yields

$$d(\phi, D, p) = d(\phi, D \backslash K, p) + d(\phi, \emptyset, p) = d(\phi, D \backslash K, p).$$

(3) Let $D_1^* = D \backslash D_1$. Since $\bar{D} \backslash (D_1 \cup D_1^*) = \partial D$, we can apply (II) to D_1 and D_1^*:

$$d(\phi, D, p) = d(\phi_1, D_1, p) + d(\phi, D_1^*, p).$$

Proceeding inductively we obtain the result.

Theorem 5.1.4 (1) *If $p \notin \overline{\phi(D)}$, then $d(\phi, D, p) = 0$.*

(2) *If $\phi(\bar{D})$ is closed and $d(\phi, D, p) \neq 0$, then there is $x \in D$ with $\phi(x) = p$.*

Proof (1) This follows from (II) with $D_1 = D_2 = \emptyset$ (noting that $\overline{\phi(\bar{D})} = \overline{\phi(D)}$).

(2) If no solution of $\phi(x) = p$ exists in $D, p \notin \phi(D)$; also $p \notin \overline{\phi(\partial D)}$ by hypothesis. Since $\phi(\bar{D})$ is closed, $p \notin \phi(D) \cup \overline{\phi(\partial D)} = \overline{\phi(D)}$. By (1), $d(\phi, D, p) = 0$. The result follows from this contradiction.

Theorem 5.1.5 (1) *For all $q \in X$, $d(\phi, D, p) = d(\phi - q, D, p - q)$.*

(2) *$d(\phi, D, .)$ is constant on components of $\mathscr{C}\overline{\phi(\partial D)}$.*

Proof (1) Since $p \notin \overline{\phi(\partial D)}$, we also have $p - q \notin \overline{(\phi - q)(\partial D)}$. By axiom (IV),

$$\begin{aligned} d(\phi - q, D, p - q) &= d(\phi - q - (p - q), D, 0) = d(\phi - p, D, 0) \\ &= d(\phi, D, p). \end{aligned}$$

(2) We suppose that $p \notin \overline{\phi(\partial D)}$; let U be a ball centre p and not meeting $\overline{\phi(\partial D)}$. Take $q \in U$, and let $p(t) = tq + (1-t)p$; then it can be seen that

$$p(t) \notin \overline{\phi(\partial D)} \quad (0 \leqslant t \leqslant 1).$$

Axiom (III′) yields

$$d(\phi, D, p) = d(\phi, D, q).$$

Hence $d(\phi, D, .)$ is a continuous mapping of $\mathscr{C}\overline{\phi(\partial D)}$ into the integers, and so is constant on components of $\mathscr{C}\overline{\phi(\partial D)}$.

The next result tells us that $d(\phi, D, p)$ is continuous in ϕ.

Theorem 5.1.6 *Suppose that $d(\phi, D, p)$ is defined and that ϕ has a convex neighbourhood in $M(D)$. There is a neighbourhood $U(\phi)$ of ϕ in $M(D)$ such that if $\psi \in U(\phi)$, then $d(\psi, D, p)$ is defined and is equal to $d(\phi, D, p)$.*

Proof Let $W(\phi)$ be a convex neighbourhood of ϕ in $M(D)$, and let

$$\rho(p, \overline{\phi(\partial D)}) = r.$$

Define $$U(\phi) = W(\phi) \cap \{\psi; \| \psi - \phi \| < r\}.$$

For $\psi \in U(\phi)$ consider the homotopy $h(t) = t\psi + (1-t)\phi$. Since $W(\phi)$ is convex, $h(t) \in M(D)$ for $t \in [0,1]$; also $h: [0,1] \to M(D)$ is continuous. Now if $x \in \bar{D}$, we have

$$h(t)(x) = \phi(x) + t(\psi(x) - \phi(x));$$

it follows that

$$\rho(p, \overline{h(t)(\partial D)}) > 0 \quad (0 \leqslant t \leqslant 1).$$

Axiom (III) then gives the result.

Corollary 5.1.7 *If $M(D)$ itself is convex, we may take $U(\phi)$ to be*

$$\{\psi; \| \phi - \psi \| < r\}.$$

Corollary 5.1.8 *Suppose that $M(D)$ is convex. Then to verify axiom (III) it is necessary only to consider homotopies $t \mapsto t\phi + (1-t)\psi$ with*

$$\|\phi(x)-\psi(x)\|<\rho(p,\overline{\phi(\partial D)})\quad (x\in\bar{D}).$$

Proof If axiom (III) holds for the particular homotopies mentioned, the proof of Theorem 5.1.6 is unchanged. Suppose that $h:[0,1]\to M(D)$ is continuous and $p\notin\overline{h(t)(\partial D)}$. Then $d(h(t),D,p)$ is defined for $0\leqslant t\leqslant 1$ and, by Theorem 5.1.6, is a continuous function of t. Hence $d(h(t),D,p)$ is independent of $t\in[0,1]$.

Our next result is a version of 'boundary dependence' (compare Theorem 2.1.4).

Theorem 5.1.9 *Suppose that $\phi,\psi\in M(D)$ are such that $\phi=\psi$ on ∂D and $p\notin\overline{\phi(\partial D)}$. If $t\phi+(1-t)\psi\in M(D)$ for $t\in[0,1]$, and $(\phi-\psi)(\bar{D})$ is bounded, then $d(\phi,D,p)=d(\psi,D,p)$.*

Proof Consider the homotopy $h(t)=t\phi+(1-t)\psi$. By hypothesis, h maps $[0,1]$ into $M(D)$. Since $(\phi-\psi)(\bar{D})$ is bounded, h is continuous. Now $\phi=\psi$ on ∂D, so $h(t)(\partial D)=\phi(\partial D)$; hence $p\notin\overline{h(t)(\partial D)}$ for $0\leqslant t\leqslant 1$. The result follows from axiom (III).

Finally in this section we come to a general fixed point theorem, which gives as special cases the theorems of Brouwer, Schauder, Tychonoff and others. Alternative versions may be obtained, just as in Chapter 4.

Theorem 5.1.10 *Suppose that $M(\omega)$ is an admissible class of mappings for X and that d is a topological degree for $M(\omega)$. Let $D\in\omega$ be such that $\phi(\bar{D})$ is bounded and $0\in D$; define $h(t)=I-t\phi$. If (a) $h(t)\in M(D)$ for $0\leqslant t\leqslant 1$, (b) the sets $(I-\phi)(\bar{D})$, $h(t)(\partial D)$ are all closed, and*

$$(c)\qquad \phi(x)\neq mx\quad (x\in\partial D,\ m>1),$$

then ϕ has a fixed point in $\bar{D}$.

Proof We suppose that $0\notin h(1)(\partial D)$, for otherwise the result is trivially true. Condition (c) then implies that

$$0\notin h(t)(\partial D)\quad (0\leqslant t\leqslant 1).$$

Since $\phi(\bar{D})$ is bounded, h is a continuous mapping of $[0,1]$ into $M(D)$. Because $h(t)(\partial D)$ is closed for $0\leqslant t\leqslant 1$, we may apply axiom

(III) to $h(t)$, giving $d(I-\phi, D, 0) = d(I, D, 0)$. By axiom (I), $d(I, D, 0) = 1$; since $(I-\phi)(\bar{D})$ is closed, the result follows from Theorem 5.1.4(2).

Remarks (1) The last few results are simplified if every $\phi \in M(D)$ is a closed mapping (that is, maps closed sets onto closed sets) and $M(D)$ is convex; fortunately, this is usually the case.

(2) It appears to be presently unknown whether a 'multiplication theorem' holds in this setting.

The theory given here can easily be taken over to the case when X is a general topological vector space. Very few specific properties of such spaces are required: indeed, only in Theorems 5.1.6 and 5.1.9 would any appreciable differences appear.

A theory of degree has been developed for topological vector spaces which are Hausdorff and locally convex (Schwartz, 1969; Nagumo, 1951b)–and these restrictions are common in work on topological vector spaces. It turns out, however, that comparatively little increase in generality is obtained by ceasing to deal exclusively with normed spaces. The reason is the following. A mapping $h:[0,1] \to M(D)$ is continuous only if $[h(t)-h(s)](\bar{D})$ is contained in a given neighbourhood of the origin when $|t-s|$ is sufficiently small. If we write

$$h(t)(x) - h(s)(x) = (t-s)\chi(x; t, s) \quad (t \neq s),$$

we must therefore have that $\chi(D; t, s)$ is bounded if $|t-s|$ is small. Now a Hausdorff, locally convex topological vector space containing a non-empty, bounded, open set, is necessarily normable (Robertson and Robertson, 1966, p. 45). If we wish to consider homotopies for which $\chi(D; t, s)$ can have non-empty interior, we may as well, therefore, stick to normed spaces.

5.2 General theory

Let us now suppose that (*a*) X is a normed space, (*b*) ω is the collection of bounded, open subsets of X, and (*c*) for all $D \in \omega$, $M(D)$ is a convex set all of whose members are bounded and proper.

Recall that ϕ is *bounded* if $\phi(A)$ is bounded whenever A is bounded, and ϕ is *proper* if $\phi^{-1}(K)$ is compact whenever K is compact. That ϕ is proper implies that ϕ is *closed* (that is, $\phi(A)$ is closed whenever A is closed). To see this, suppose that A is closed and $\phi(x_n) \to y$ $(x_n \in A)$; the set $S = \{y, \phi(x_1), \phi(x_2), \dots\}$ is compact, whence $\phi^{-1}(S)$ is compact. Therefore (x_n) has a convergent subsequence: say $x_{n_k} \to x$; since A is closed, $x \in A$. Now ϕ is continuous, so $y = \phi(x) \in \phi(A)$.

The reader will observe that what we say is valid more generally, but these hypotheses effect useful simplifications, some of which we list below.

(*a*) Since ∂D is closed, the degree $d(\phi, D, p)$ is defined when $p \notin \phi(\partial D)$; correspondingly, the condition in axiom (III) becomes $p \notin h(t)(\partial D)$.

(*b*) Theorem 5.1.3(3) holds when D is the disjoint union of countably many sets D_i in ω (the summation is finite, by Theorem 5.1.4 and the compactness of $\phi^{-1}(p)$).

(*c*) Theorem 5.1.4 states that $d(\phi, D, p) = 0$ if $p \notin \phi(\bar{D})$, and conversely that $p \in \phi(D)$ when $d(\phi, D, p) \neq 0$.

(*d*) The hypotheses of Theorems 5.1.6 and 5.1.9 are automatically satisfied, so that these results hold without restriction.

(*e*) In the fixed point theorem (Theorem 5.1.10), $\phi(\bar{D})$ is bounded, and conditions (*a*) and (*b*) can be removed.

As before $K_1(\bar{D})$ denotes the collection of mappings $\bar{D} \to X$ which are compact perturbations of the identity. The following is well known and will be needed later.

Lemma 5.2.1 *If $\phi \in K_1(\bar{D})$, then ϕ is bounded, closed and proper.*

A mapping $\phi \in C(\bar{D})$ is said to be *Fréchet-differentiable* (or just *differentiable*) at $x_0 \in D$ if there is a linear mapping $\phi'(x_0): X \to X$ such that

$$\phi(x_0 + h) - \phi(x_0) - \phi'(x_0)h = o(\|h\|) \quad (h \to 0).$$

We shall say that ϕ is differentiable in $\bar{D}$ if there is an extension $\tilde{\phi}$ of ϕ to a neighbourhood of $\bar{D}$ in which $\tilde{\phi}$ is everywhere differentiable. Let $C^1(\bar{D})$ denote the vector space of mappings $\bar{D} \to X$ which are continuously differentiable in $\bar{D}$. Also let $L(X)$ denote the space of continuous linear mappings of X into itself, and $GL(X)$ the group of isomorphisms of X onto itself.

Suppose that $\phi \in M(D)$ and $x_0 \in D$ is an isolated p-point of ϕ (that is, an isolated solution of $\phi(x) = p$). Then, for all sufficiently small neighbourhoods U of $x_0, \phi \in M(U)$ and $d(\phi, U, p)$ is defined; moreover, by axiom (II), $d(\phi, U, p)$ is independent of U.

Definition 5.2.2 The *index* $i(\phi, x_0, p)$ of the isolated p-point x_0 of ϕ is $d(\phi, U, p)$, where U is any neighbourhood of x_0 in D containing no other solution of $\phi(x) = p$.

Theorem 5.2.3 *Let $\phi \in C^1(\bar{D}) \cap M(D)$; suppose that $x_0 \in D, \phi(x_0) = p$ and $\phi'(x_0) \in GL(X) \cap M(D)$. Then x_0 is an isolated p-point of ϕ and, if $y_0 = \phi'(x_0)x_0$,*

$$i(\phi, x_0, p) = i(\phi'(x_0) - y_0, x_0, 0).$$

Proof Since $\phi'(x_0) \in GL(X)$, there is $\alpha > 0$ such that

$$\|\phi'(x_0)x\| \geqslant \alpha \|x\| \quad (x \in X). \tag{5.2.1}$$

Let $U \subset D$ be a neighbourhood of x_0 such that

$$\|\phi(x) - p - \phi'(x_0)(x - x_0)\| \leqslant \tfrac{1}{3}\alpha \|x - x_0\| \quad (x \in U). \tag{5.2.2}$$

It is clear from (5.2.1) and (5.2.2) that x_0 is an isolated p-point of ϕ. Define

$$h(t)(x) = (1 - t)\phi(x) + t\phi'(x_0)(x - x_0) \quad (0 \leqslant t \leqslant 1, x \in \bar{U})$$

and $\quad p(t) = (1 - t)p \quad (0 \leqslant t \leqslant 1).$

Then
$$\begin{aligned}&\|h(t)(x) - p(t)\| \\ &\geqslant \|\phi'(x_0)(x - x_0)\| - (1 - t)\|\phi(x) - p - \phi'(x_0)(x - x_0)\| \\ &\geqslant \tfrac{2}{3}\alpha \|x - x_0\|.\end{aligned}$$

So $p(t) \notin h(t)(\partial U)$, whence, by axiom (III′),

$$d(\phi, U, p) = d(\psi, U, 0),$$

where $\psi(x) = \phi'(x_0)(x - x_0) = \phi'(x_0)x - y_0$. This is the required result.

The value of this theorem is that it reduces the calculation of the degree of a mapping to that of the degree of a linear mapping. If $\phi \in C^1(\bar{D})$ and p is such that $\phi'(x) \in GL(X)$ for all $x \in \phi^{-1}(p)$, then $\phi^{-1}(p)$ is a finite set if ϕ is proper. In that case, we see from Theorem 5.1.3(3) that

$$d(\phi, D, p) = \sum_{\phi(x)=p} i(\phi, x, p).$$

If $\phi \notin C^1(\bar{D})$, it is often possible to find ψ arbitrarily close to ϕ with $\psi \in C^1(\bar{D})$ and $\psi'(x) \in GL(X)$ for $x \in \psi^{-1}(p)$. Then, by Theorem 5.1.6, $d(\phi, D, p) = d(\psi, D, p)$. Such an approximation is always possible when X is finite dimensional, for we can use Sard's theorem.

Suppose now that there exists a topological degree d for $M(\omega)$, but that D is not an open subset of X itself. If, however, D is an open subset of a subspace Y of X, it is clearly desirable to define a degree for suitable mappings relative to D, induced by d. Let $\omega_Y = \{Y \cap A\,; A \in \omega\}$; if $D \in \omega_Y$ we wish to define a degree $d_Y(\phi, D, p)$ for suitable mappings ϕ of $\bar{D}$ into Y. We shall suppose that Y is a topological direct summand of X: let $X = Y \oplus Z$. For $S \subset X$, write $S^{(1)}$ and $S^{(2)}$ for the projections of S on Y and Z, respectively; if $f: S^{(1)} \to Y$ and $g: S^{(2)} \to Z$, we define $f \oplus g: S \to X$ by

$$f \oplus g(x) = f(y) + g(z) \quad (x = y + z). \tag{5.2.3}$$

If f and g are continuous, then, because $X = Y \oplus Z$ is a topological direct sum, $f \oplus g$ is also continuous.

Let U be a bounded neighbourhood of 0 in Z. Provided that $p \in Y$ and $\phi \oplus I \in M(D \oplus U)$, we define

$$d_Y(\phi, D, p) = d(\phi \oplus I, D \oplus U, p). \tag{5.2.4}$$

The p-points of $\phi \oplus I$ lie in D, so the right hand side of (5.2.4) is certainly defined if $p \notin \phi(\partial_Y D)$ – where ∂_Y is, of course, the boundary operator in Y. We have made two hypotheses: the requirement that $\phi \oplus I \in M(D \oplus U)$ is very mild, and will almost certainly be satisfied in applications (for example, if $\phi \in K_1(\bar{D})$, and $M(D \oplus U) = K_1(\overline{D \oplus U})$); that Y is a topological direct summand in X is not unduly restrictive and it is satisfied if Y is finite dimensional.

Theorem 5.2.4 *d_Y as defined in (5.2.4) is a degree for*

$$M(\omega_Y) = \{\phi ; \phi \oplus I \in M(D \oplus U), D \in \omega_Y\}.$$

Proof The four axioms must be checked in turn; we leave the details to the reader.

Remarks The degree d_Y is independent of the particular complement of Y that is chosen and of the choice of U. If V is a subspace of Y, we also have

$$(d_Y)_V = d_V.$$

We now show how to calculate from the axioms the degree of linear mappings when the ambient space X is finite dimensional. By Theorem 5.2.3 and the remarks immediately following it, we are then (theoretically) in a good position to determine the degree of all relevant mappings. Thus, in particular, the axioms determine d uniquely when X is finite dimensional. The details will be given in Section 5.3. With Theorems 5.2.3 and 5.2.6 (to follow), we shall have virtually recovered the original definition of the Brouwer degree.

Theorem 5.2.5 *Suppose that* $\dim X = n < \infty$ *and that* $D \in \omega'$. *If* $0 \in D$, *then (provided $M(D)$ is large enough)*

$$d(-I, D, 0) = (-1)^n.$$

Note The significance of the hypothesis 'provided $M(D)$ is large enough' is that the proof requires $M(D)$ to contain a few simple, but non-trivial, functions – the details will be seen in the proof. Recall that we are in this section requiring $M(D)$ to be convex.

Proof (i) Suppose first that n is even. Define

$$T: (x_1, x_2, \dots, x_{n-1}, x_n) \mapsto (x_2, -x_1, \dots, x_n, -x_{n-1}).$$

Clearly $T \in GL(X)$ and $T^2 = -I$. Consider the homotopies

$$h_1(t) = tT + (1-t)(-I), \quad h_2(t) = tT + (1-t)I \quad (0 \leqslant t \leqslant 1).$$

It is easily checked that for $0 \leqslant t \leqslant 1$,

$$0 \notin h_1(t)(\partial D) \quad \text{and} \quad 0 \notin h_2(t)(\partial D).$$

Provided $T \in M(D)$, therefore, $d(-I, D, 0) = d(T, D, 0) = d(I, D, 0)$, by axiom (III). But, by axiom (I), $d(I, D, 0) = 1$ – whence the result.

(ii) Next, suppose that n is odd and $n > 1$. Define

$$T_1 : (x_1, x_2, \ldots, x_{n-1}, x_n) \mapsto (-x_1, x_3, -x_2, \ldots, x_n, -x_{n-1}).$$

and $$\phi : (x_1, x_2, \ldots, x_{n-1}, x_n) \mapsto (-x_1, x_2, x_3, \ldots, x_{n-1}, x_n).$$

An argument similar to that in (i) shows that

$$d(-I, D, 0) = d(\phi, D, 0),$$

for both $-I$ and ϕ are homotopic to T_1.

(iii) It remains to calculate $d(\phi, D, 0)$; we shall show that it is just $-d(I, D, 0)$; we now admit the value $n = 1$. We suppose that an orthogonal basis has been so chosen for X that

$$D \supset D_1 = (-1, 3)^n.$$

Let

$$D_2 = (-1, 1) \times (-1, 3)^{n-1} \quad \text{and} \quad D_3 = (1, 3) \times (-1, 3)^{n-1}.$$

Define $\phi_1(x) = (|x_1 - 1| - 1, x_2, \ldots, x_n)$, $\phi_2(x) = (1, x_2, \ldots, x_n)$ and $\phi_3(x) = (x_1 - 2, x_2, \ldots, x_n)$. These three are continuous mappings $\bar{D} \to X$; we suppose that all three are members of $M(D)$. In the following every homotopy used is of the 'straight line' kind: $t(.) + (1-t)(..)$. Now

$$\begin{aligned} d(\phi_1, D_1, 0) &= d(\phi_1, D_2, 0) + d(\phi_1, D_3, 0) \quad &\text{(Theorem 5.1.3)} \\ &= d(\phi, D_2, 0) + d(\phi_3, D_3, 0) \quad &\text{(axiom (III))}. \end{aligned}$$

If we write $a = (2, 0, \ldots, 0)$,

$$\begin{aligned} d(\phi_3, D_3, 0) &= d(I, D_3, a) \quad &\text{(Theorem 5.1.5(1))} \\ &= 1 \quad &\text{(axiom (I))}. \end{aligned}$$

But $$\begin{aligned} d(\phi_1, D_1, 0) &= d(\phi_2, D_1, 0) \quad &\text{(axiom (III))} \\ &= 0 \quad &\text{(Theorem 5.1.4(2))}. \end{aligned}$$

Therefore $d(\phi, D_2, 0) = -1$. By Theorem 5.1.3(2), $d(\phi, D, 0) = d(\phi, D_2, 0)$; it follows that $d(\phi, D, 0) = -1$.

(iv) Putting the results of (ii) and (iii) together, we have $d(-I, D, 0) = -1$ if n is odd. With (i), this completes the proof of the theorem.

Theorem 5.2.6 *Suppose that X is finite dimensional and $T \in GL(X)$. Let N be the linear space spanned by the set*

$$\{x ; (T - \lambda I)^k x = 0 \text{ for some } k > 0 \text{ and } \lambda < 0\}.$$

If $\dim N = n$ *and* $0 \in D$, *then* $d(T, D, 0) = (-1)^n$.

Note n is the number of real, negative eigenvalues of T, counting multiplicity.

Proof Suppose that $n > 0$; let $\lambda_1, \ldots, \lambda_k$ be the distinct negative eigenvalues of T. It is known from the theory of vector spaces that there is a decomposition

$$X = N_1 \oplus N_2 \oplus \ldots \oplus N_k \oplus M,$$

where M and all the N_i are invariant under T. The eigenvalues of $T|_M$ are those of T other than $\lambda_1, \ldots, \lambda_k$, and λ_i is the only eigenvalue of $T|_{N_i}$; $\dim N_i = m_i$ is the algebraic multiplicity of the eigenvalue λ_i. Now

$$N = N_1 \oplus N_2 \oplus \ldots \oplus N_k.$$

Define the homotopy

$$h(t) = (1 - t)(T_1 \oplus T_2) + t(-I_1 \oplus I_2) \quad (0 \leqslant t \leqslant 1),$$

where the subscripts 1, 2 denote restriction to N, M respectively. Since T_1 has no positive eigenvalues and T_2 no negative ones, $0 \notin h(t)(\bar{D})$ for $0 \leqslant t \leqslant 1$. Therefore

$$d(T, D, 0) = d_N(-I, D \cap N, 0),$$

in the previous notation. By Theorem 5.2.5, $d_N(-I, D \cap N, 0)$ is $(-1)^n$, whence the result. The same argument also works (in a simplified form) when $n = 0$.

5.3 The uniqueness of the Leray–Schauder degree

In Chapter 4 we defined a degree, the Leray–Schauder degree, when X was a normed space, $\omega = \beta$, the bounded, open subsets of X, and $M(D) = K_1(\bar{D})$. We proceed in this section to show that the Leray–Schauder degree is unique in the sense that it is the only possible definition of degree in this context.

Let d be a topological degree for $K_1(\beta) = \{K_1(\bar{D}); D \in \beta\}$. We start with $p \in X$, $D \in \beta$ and $\phi \in K_1(\bar{D})$, and reduce the evaluation of $d(\phi, D, p)$ to successively simpler cases.

(i) Axiom (IV) shows that $d(\phi, D, p) = d(\phi - p, D, 0)$; d is therefore determined if we consider only $p = 0$.

(ii) Let $K_2(\bar{D})$ be the subset of $K_1(\bar{D})$ consisting of those $\phi = I - f$ for which f has finite dimensional range. We saw in Theorem 4.2.2 that $K_2(\bar{D})$ is dense in $K_1(\bar{D})$ – recall that $K_1(\bar{D})$ has the topology of uniform convergence. It then follows from Theorem 5.1.6 that $d(\phi, D, 0) = d(\psi, D, 0)$, where $\psi \in K_2(\bar{D})$ is sufficiently near to ϕ. So d is determined if we consider only $\phi \in K_2(\bar{D})$.

(iii) The next stage is the longest, and utilises the ideas and notation of Section 5.2. Suppose that $\phi = I - f \in K_2(\bar{D})$ and $f(\bar{D}) \subset F$, a finite dimensional subspace of X. F is certainly a topological direct summand of X: say $X = F \oplus G$ (see, for example, Dieudonné (1960, p. 108)). If $D \cap F = \emptyset$, then, by considering the zeros of ϕ, we see that $d(\phi, D, 0) = 0$; we therefore assume henceforth that $D \cap F \neq \emptyset$. By Tietze's theorem, there is a continuous mapping $\hat{\phi}: F \to F$ such that $\hat{\phi} = \phi$ on $\bar{D} \cap F$. Let U be a (fixed) bounded neighbourhood of 0 in G; define

$$h(t) = t\phi + (1 - t)(\hat{\phi} \oplus I_G) \quad (0 \leqslant t \leqslant 1).$$

Write $x = y + z$, with $y \in F$ and $z \in G$; we have

$$(I - \hat{\phi} \oplus I_G)(x) = (I - \hat{\phi})(y).$$

We see that h maps $[0, 1]$ into $K_1(\bar{D})$, and moreover is continuous. If $h(t)(x) = 0$, then

$$t(y - f(x)) + (1 - t)\hat{\phi}(y) + z = 0. \tag{5.3.1}$$

But $y - f(x) \in F$ and $\hat{\phi}(y) \in F$, so we must have $z = 0$; then (5.3.1) becomes

$$t\phi(y)+(1-t)\hat{\phi}(y)=0.$$

There can therefore be no solution of (5.3.1) in ∂D – it is, of course, implied throughout that $0 \notin \phi(\partial D)$. By homotopy invariance, we have

$$d(\phi, D, 0) = d(\hat{\phi} \oplus I_G, D, 0).$$

The equation (5.3.1) also shows that $\hat{\phi} \oplus I_G$ can have no zeros in D outside $D_F = D \cap F$; the excision property (Theorem 5.1.3(2)) implies that

$$d(\hat{\phi} \oplus I_G, D, 0) = d(\hat{\phi} \oplus I_G, D_F \oplus U, 0).$$

But $d(\hat{\phi} \oplus I_G, D_F \oplus U, 0) = d_F(\hat{\phi}, D_F, 0)$ (see (5.2.4)). The evaluation of $d(\phi, D, 0)$ has thus been reduced to that of $d_F(\hat{\phi}, D_F, 0)$. This degree is of the form $d(f, \Delta, 0)$, where Δ is a bounded, open subset of a finite dimensional space and $f \in C(\bar{\Delta})$. Following Theorem 5.2.4, we remarked that the axioms determine such degrees uniquely. We now prove this in detail. We suppose henceforth that $\dim X < \infty$ and that $\phi \in C(\bar{D})$.

(iv) Having reduced the problem to one in finite dimensional space, we have Sard's theorem available to us. It is possible to choose a C^1 function ψ in any neighbourhood of ϕ in $C(\bar{D})$ such that 0 is a regular value of ψ. By Theorem 5.1.6, $d(\phi, D, 0) = d(\psi, D, 0)$ if ψ is sufficiently near ϕ. So d is determined if we consider only functions ϕ which are C^1 and for which 0 is not a critical value.

(v) Under our latest hypotheses (that 0 is a regular value of the C^1 function ϕ), $\phi^{-1}(0)$ is a finite set – $\{x_1, \ldots, x_k\}$, say. By the comments following Theorem 5.2.3,

$$d(\phi, D, 0) = \sum_{j=1}^{k} i(\phi, x_j, 0).$$

By Theorem 5.2.3,

$$i(\phi, x_j, 0) = i(T_j - y_j, x_j, 0) \quad (j = 1, \ldots, k),$$

where T_j is the linear map $x \mapsto \phi'(x_j)x$ and $y_j = T_j(x_j)$. By hypothesis, $T_j \in GL(X)$ $(j = 1, \ldots, k)$.

(vi) It remains to evaluate the indices $i(T_j - y_j, x_j, 0)$. Consider a fixed j – say $j=1$, for convenience. Since T is an isomorphism, x_1 is the only solution of $T_1(x)=y_1$, so that $i(T_1-y_1, x_1, 0)=d(T_1-y_1, W, 0)$ for any bounded neighbourhood W of x_1. So choose W that it contains the entire line segment $\{tT_1^{-1}(y_1); 0\leqslant t\leqslant 1\}$. Let

$$h(t)(x)=(1-t)T_1(x)+t(T_1(x)-y_1) \quad (0\leqslant t\leqslant 1, x\in \bar{W}).$$

By the choice of $W, 0\notin h(t)(\partial W)$ for $0\leqslant t\leqslant 1$. Axiom (III) then tells us that

$$d(T_1-y_1, W, 0)=d(T_1, W, 0).$$

Now $d(T_1, W, 0)$ can be calculated using Theorem 5.2.6. In this way, each of the indices $i(T_j-y_j, x_j, 0)$ can be found.

The process is now complete; we have proved the following.

Theorem 5.3.1 *Suppose that X is a normed space, D is a bounded, open subset of X, and d is a topological degree for $K_1(\bar{D})$. If $\phi\in K_1(\bar{D})$ and $p\in X\setminus\phi(\partial D)$, then $d(\phi, D, p)$ is necessarily equal to the Leray-Schauder degree of ϕ at p relative to D.*

We also have the two following results.

Theorem 5.3.2 *Suppose that X is a finite dimensional normed space, D is a bounded, open subset of X, and d is a topological degree for $C(\bar{D})$. If $\phi\in C(\bar{D})$ and $p\in X\setminus\phi(\partial D)$, then $d(\phi, D, p)$ must be equal to the Brouwer degree of ϕ at p relative to D.*

Theorem 5.3.3 *If d is a topological degree defined for an admissible class of mappings $\{I-\phi\}$, where ϕ can be approximated by compact mappings, then d is uniquely determined.*

6
Condensing maps and k-set contractions

In this chapter we extend the class of mappings for which it is possible to define a topological degree. We continue to suppose that the ambient space X is an infinite dimensional normed space, and, moreover, that it is *complete* (that is, X is a Banach space). By Theorem 5.3.1, the degree we define necessarily coincides with the Leray–Schauder degree for compact perturbations of the identity.

The recent development of non-linear functional analysis has placed considerable emphasis on the existence of solutions of various kinds of operator equations in Banach spaces. Several classes of mappings containing the compact mappings have been introduced (for example, the A-proper and accretive mappings). The first of these that we investigate is the class of k-set contractions. Degree for such mappings was defined by Nussbaum (1969, 1971*a*, 1972) using the fixed point index of compact mappings; the treatment given here is somewhat different.

6.1 Measure of non-compactness

In this section, the setting initially is a complete metric space (Y, σ).

Definition 6.1.1 Let S be a bounded subset of Y. Define $\Delta(S) \subset \mathbf{R}$ to be the set of $\delta > 0$ such that S can be covered with a finite number of sets of diameter less than δ. The *measure of non-compactness* of S, $\alpha(S)$, is $\inf \Delta(S)$.

The notion of the measure of non-compactness was introduced by Kuratowski (1930); the motivation was the following result.

Theorem 6.1.2 *Suppose that (A_n) is a decreasing sequence of non-empty closed sets such that $\alpha_n = \alpha(A_n) \to 0$ as $n \to \infty$. Then $A = \bigcap_{n=1}^{\infty} A_n$ is non-empty and compact.*

Proof We denote the diameter of a bounded set S by $\delta(S)$. Let

$$A_i = A_{i1} \cup \ldots \cup A_{im_i} \quad (i = 1, 2, 3, \ldots),$$

where $\delta(A_{im_j}) < \alpha_i + i^{-1} (j = 1, \ldots, i)$. We shall show that every sequence (p_n), so chosen that $p_n \in A_n$, has a convergent subsequence; the limit must lie in A, whence A is non-empty. The same argument shows that every sequence in A has a convergent subsequence, proving that A is compact.

Consider $S = \{p_1, p_2, \ldots\}$, with $p_n \in A_n$; we seek a subsequence $(p_{k_1}, p_{k_2}, \ldots)$ which converges. Since $S \subset A_1$, at least one of $A_{11}, \ldots, A_{1m_1}$ contains an infinite subset of S – say

$$S_1 = \{p_{l_1}, p_{l_2}, \ldots\} \subset A_{11} \quad (1 < l_1 < l_2 < \ldots).$$

Take $k_1 = 1$ and $k_2 = l_1$. Since $S_1 \subset A_2$, at least one of $A_{21}, \ldots, A_{2m_2}$ contains an infinite subset S_2 of S_1 – say

$$S_2 = \{p_{\lambda_1}, p_{\lambda_2}, \ldots\} \subset A_{21} \quad (l_1 < \lambda_1 < \lambda_2 < \ldots).$$

Take $k_3 = \lambda_1$. In this way a sequence $\{p_{k_1}, p_{k_2}, \ldots\}$ is inductively defined. We have

$$\sigma(p_{k_n}, p_{k_{n+m}}) < \delta(A_{n-1,1}) < \alpha_{n-1} + (n-1)^{-1}.$$

So $\{p_{k_n}\}$ is a Cauchy sequence in the complete space (Y, σ), and is therefore convergent. This completes the proof.

The main properties of the measure of non-compactness of sets are listed in the next theorem.

Theorem 6.1.3 *Let S and T be bounded subsets of Y. Then* (1) $S \subset T$ *implies* $\alpha(S) \leqslant \alpha(T)$, (2) $\alpha(\bar{S}) = \alpha(S)$, (3) $\alpha(S) = 0$ *if and only if S is relatively compact,* (4) $\alpha(S \cup T) = \max(\alpha(S), \alpha(T))$.

Proof If $S \subset T$, then any cover of T is a cover of S; hence $\alpha(S) \leqslant \alpha(T)$, which is (1). From this it follows in particular that $\alpha(S) \leqslant \alpha(\bar{S})$. On the other hand, if $\{S_1, \ldots, S_N\}$ is a cover of S, then $\{\bar{S}_1, \ldots, \bar{S}_N\}$

is a cover of $\bar{S}$; since $\delta(\bar{S}_i) = \delta(S_i)$ $(i = 1, \ldots, N)$, we have $\alpha(\bar{S}) \leqslant \alpha(S)$. Part (2) of the theorem follows. As for (3), it is obvious that if S is relatively compact, then $\bar{S}$ can be covered with a finite number of sets of arbitrarily small diameter; hence $\alpha(\bar{S}) = 0$, and, by (2), $\alpha(S) = 0$. Conversely, suppose $\alpha(S) = 0$. In Theorem 6.1.2 let $A_n = \bar{S}$ for all n; it follows that $\bar{S} = \bigcap A_n$ is compact. We leave (4) as a simple exercise for the reader.

For the next three results, Y is a Banach space $(X, \|.\|)$. It is necessary to compare the measure of non-compactness of the convex hull co S of a set S with that of S itself. Theorem 6.1.5 was proved by Darbo (1955).

Lemma 6.1.4 *Let S be a bounded subset of X; then $\delta(S) = \delta(\text{co } S)$.*

Proof Since $S \subset \text{co } S$, certainly $\delta(S) \leqslant \delta(\text{co } S)$. Suppose, if possible, that $\delta(S) < \delta(\text{co } S)$; there must then be points x, y in co S with $\|x - y\| > \delta(S)$. Let S_1 be the ball of radius $\delta(S)$ with centre x: $S_1 = B(x, \delta(S))$. If $S \subset S_1$, then $S_1 \cap \text{co } S$ is a convex set containing S, whence $S_1 \cap \text{co } S = \text{co } S$. But $y \in (\text{co } S)\backslash S_1$, a contradiction. On the other hand, if there is $z \in S\backslash S_1$, let $S_2 = B(z, \delta(S))$. Then $S_2 \supset S$, whence $S_2 \supset \text{co } S$; but $x \in (\text{co } S)\backslash S_2$, another contradiction. We conclude that $\delta(S) = \delta(\text{co } S)$.

Theorem 6.1.5 *If $S \subset X$ is bounded, then $\alpha(\overline{\text{co}}\ S) = \alpha(S)$.*

Proof Since $\overline{\text{co}}\ S = \overline{\text{co } S}$, it is necessary only to prove that $\alpha(S) = \alpha(\text{co } S)$. Now $S \subset \text{co } S$, so that $\alpha(S) \leqslant \alpha(\text{co } S)$; we must therefore show that $\alpha(S) \geqslant \alpha(\text{co } S)$. This we do in four steps.

(*a*) Take $\varepsilon > \alpha(S)$, and suppose that $X_1, \ldots, X_n$ cover S, with $\delta(X_r) < \varepsilon$ $(r = 1, \ldots, n)$. So choose ε_1 that $\delta(X_r) \leqslant \varepsilon_1 < \varepsilon$ for all r; by Lemma 6.1.4, $\delta(\text{co } X_r) \leqslant \varepsilon_1$. Define a subset A of $\mathbf{R}^n$ by

$$A = \{\lambda = (\lambda_1, \ldots, \lambda_n);\ \sum_{i=1}^{n} \lambda_i = 1,\ \lambda_i \geqslant 0\ (i = 1, \ldots, n)\}.$$

Corresponding to $\lambda \in A$, define a subset $Y(\lambda)$ of X:

$$Y(\lambda) = \{x;\ x = \sum_{i=1}^{n} \lambda_i x_i,\ x_i \in \text{co } X_i\}.$$

(*b*) Suppose that $x, y \in Y(\lambda)$; let $x = \sum \lambda_i x_i$ and $y = \sum \lambda_i y_i$ (where $x_i, y_i \in \text{co } X_i$). Then

$$\|x - y\| \leqslant \sum \lambda_i \|x_i - y_i\| \leqslant \sum \lambda_i \delta(\text{co } X_i) \leqslant \varepsilon_1 \sum \lambda_i = \varepsilon_1 .$$

Therefore $\delta(Y(\lambda)) \leqslant \varepsilon_1$. It is clear that $\{Y(\lambda); \lambda \in A\}$ is a cover of S.

(*c*) Write $$Y(A) = \bigcup \{Y(\lambda); \lambda \in A\}.$$

We say that $Y(A)$ is a convex set containing S and contained in co S. For let $x, y \in Y(A)$; suppose that $x = \sum_{i=1}^{n} \lambda_i x_i$ and $y = \sum_{i=1}^{n} \mu_i y_i$, where $\lambda = (\lambda_1, \ldots, \lambda_n)$, $\mu = (\mu_1, \ldots, \mu_n)$ and $x_i, y_i \in \text{co } X_i (i = 1, \ldots, n)$. Consider a point

$$z = hx + (1 - h)y \quad (0 < h < 1);$$

it may be easily checked that $z = \sum_{i=1}^{n} \theta_i z_i$, where

$$z_i = (h\lambda_i x_i + (1 - h)\mu_i y_i)/(h\lambda_i + (1 - h)\mu_i)$$

and $$\theta_i = h\lambda_i + (1 - h)\mu_i .$$

It follows that $z \in Y(\theta) \subset Y(A)$. This means that $Y(A)$ is convex. It is clear that $S \subset Y(A)$ and that $Y(A) \subset \text{co } S$; therefore $Y(A) = \text{co } S$.

(*d*) For $\eta > 0$, denote the η-neighbourhood of $Y(\lambda)$ by $[Y(\lambda)]_\eta$. Given $\lambda \in A$ and $\eta > 0$, there is ε_2 such that, if $|\lambda - \mu| < \varepsilon_2$, then

$$Y(\mu) \subset [Y(\lambda)]_\eta .$$

Now A is a compact subset of $\mathbf{R}^n$, so it is covered by neighbourhoods $N_1, \ldots, N_\nu$ of finitely many points $\lambda_1, \ldots, \lambda_\nu$ of A, so chosen that

$$Y(\lambda) \subset [Y(\lambda_j)]_\eta \quad (\lambda \in N_j, j = 1, \ldots, \nu).$$

Take $\eta = \frac{1}{3}(\varepsilon - \varepsilon_1)$; then, for $j = 1, \ldots, \nu$,

$$\delta(Y(N_j)) \leqslant \delta([Y(\lambda_j)]_\eta) \leqslant \delta(Y(\lambda_j)) + 2\eta \leqslant \varepsilon_1 + 2\eta < \varepsilon.$$

Hence $$\text{co } S = Y(A) = \bigcup_{j=1}^{\nu} Y(N_j),$$

with $\delta(Y(N_j)) < \varepsilon$. This means that $\alpha(\text{co } S) \leqslant \alpha(S)$, as required. The proof is complete.

The next result is also due to Darbo (1955); the proof, which is very easy, is omitted.

Theorem 6.1.6 *Let S and T be bounded subsets of X; then*

$$\alpha(S+T) \leqslant \alpha(S) + \alpha(T).$$

Another 'measure of non-compactness' for metric spaces was introduced by Gokhberg, Goldenstein and Markus (1957); it is the ball measure of non-compactness, and is defined to be $\beta(S)$, where

$$\beta(S) = \inf\{\varepsilon > 0; S \text{ can be covered with finitely many balls of radius } \varepsilon\}.$$

The properties of α and β are very similar. It is not difficult to see that $\beta(S) \leqslant \alpha(S) \leqslant 2\beta(S)$ for any bounded set S; generally, though, little can be said about the connection between α and β. Nussbaum (1971*a*) gave the following example; it shows that $\alpha(S)$ and $\beta(S)$ may not be equal and also serves as an illustration of the foregoing.

Example Let X be an infinite dimensional Banach space, and let B, U denote the unit ball and unit sphere, respectively. It is not difficult to show that $\beta(B) = 1$. We shall prove that $\alpha(B) = 2$. By Theorem 6.1.5, $\alpha(U) = \alpha(\overline{\text{co}}\ U) = \alpha(B)$; now $\delta(U) = 2$, so certainly $\alpha(U) \leqslant 2$. If $\alpha(U) < 2$, we can write $U = T_1 \cup \ldots \cup T_n$, with $\delta(T_i) < 2$ $(i = 1, \ldots, n)$; by taking $\overline{T}_i$ instead of T_i, if necessary, we can suppose that the sets T_i are closed. Let F be an n-dimensional subspace of X; then $U \cap F$ is covered with closed sets $T_1 \cap F, \ldots, T_n \cap F$. By Corollary 3.2.9, at least one of the $T_i \cap F$ contains a pair of antipodal points; if $T_k \cap F$ contains such a pair,

$$2 \leqslant \delta(T_k \cap F) \leqslant \delta(T_k),$$

which is a contradiction. We conclude that $\alpha(U) = 2$, and hence that $\alpha(B) = 2$.

The example shows that we can have $\alpha(S) = 2\beta(S)$. That this relation does not necessarily hold is shown by taking the metric space Y to be the unit sphere U of an infinite dimensional Hilbert space, with the induced topology. Then $\alpha(U)=2$, but $\beta(U)=\sqrt{2}$. This illustrates one of the differences between the α and β measures: the value of $\beta(S)$ depends on the ambient space, while that of $\alpha(S)$ does not.

The machinery that has been introduced enables us to define classes of mappings which are in a sense 'contracting' and which contain the collection of compact mappings as proper subclasses.

Definition 6.1.7 Let (Y_1,σ_1) and (Y_2,σ_2) be metric spaces, and suppose that $\phi: Y_1 \to Y_2$ is continuous.

(*a*) ϕ is *condensing* if, for every bounded, non-compact subset S of Y_1, $\phi(S)$ is bounded and $\alpha(\phi(S)) < \alpha(S)$.

(*b*) ϕ is a *k-set contraction* if, for all bounded subsets S of $Y_1, \phi(S)$ is bounded and $\alpha(\phi(S)) \leqslant k\alpha(S)$.

(*c*) ϕ is a *strict set contraction* if it is a k-set contraction with $k<1$.

We shall describe $\phi: Y_1 \to Y_2$ as a *local condensing map* if, given $x \in Y_1$, there exists a neighbourhood U_x of x such that $\phi|_{U_x}$ is condensing; similarly, we say that ϕ is a *local k-set contraction* if, given x, there is a neighbourhood U_x such that $\phi|_{U_x}$ is a k-set contraction.

It is easily seen that a strict set contraction is necessarily condensing and that a condensing mapping is a 1-set contraction. Nussbaum (1971*a*, p. 222) gives an example of a mapping which is condensing but not a strict set contraction.

Definition 6.1.7 may be adapted to apply when the measure of non-compactness α is replaced by the ball measure of non-compactness β. The corresponding properties are described as 'ball-condensing' and 'k-ball contraction'. These have been studied by Sadovskii (1967) and Furi and Vignoli (1970).

Henceforth we restrict attention to a Banach space X.

Theorem 6.1.8 *Suppose that* Γ *is a subset of the Banach space* X.

(1) *If* $\phi:\Gamma\to X$ *is a Lipschitz mapping with constant* L, *then* ϕ *is an* L*-set contraction.*

(2) *If $\phi : \Gamma \to X$ is continuous, it is compact if and only if it is a 0-set contraction.*

Proof (1) Suppose that S is a bounded subset of Γ, and let $\alpha(S) = d$. Given $\varepsilon > 0$, there is a cover $\{S_1, \ldots, S_m\}$ of S with $\delta(S_i) \leqslant d + \varepsilon$ $(i = 1, \ldots, m)$. By hypothesis $\| \phi(x) - \phi(y) \| \leqslant L \| x - y \|$; hence $\delta(\phi(S_i)) \leqslant L\delta(S_i)$. So $\{\phi(S_1), \ldots, \phi(S_m)\}$ is a cover of $\phi(S)$ with $\delta(\phi(S_i)) \leqslant L(d + \varepsilon)$ $(i = 1, \ldots, m)$. Now ε is arbitrary, so $\alpha(\phi(S)) \leqslant Ld = L\alpha(S)$, as required.

(2) Suppose that ϕ is compact. Let $S \subset \Gamma$ be bounded; then $\phi(S)$ is relatively compact, so that $\alpha(\phi(S)) = 0$ (Theorem 6.1.3(3)). It is immediate that ϕ is a 0-set contraction. Conversely, if $\alpha(\phi(S)) = 0$ for all bounded subsets S of Γ, $\phi(S)$ is relatively compact whenever S is bounded (Theorem 6.1.3(3) again); thus ϕ is compact.

Theorem 6.1.9 *Suppose that ϕ and ψ are mappings of X into itself. If ϕ is a k_1-set contraction and ψ is a k_2-set contraction, then $\phi + \psi$ is a $(k_1 + k_2)$-set contraction, and $\phi \circ \psi$ is a $k_1 k_2$-set contraction.*

Corollary 6.1.10 *The sum of a compact mapping and a contraction mapping is a strict set contraction.*

Proofs The proofs of the results contained in the theorem are straightforward, and we omit them. The corollary is proved simply by observing that a compact mapping is a 0-set contraction and that a contraction mapping is a k-set contraction with $k < 1$ (both by Theorem 6.1.8); the theorem then gives the conclusion at once.

6.2 Degree for condensing maps

Let D be a bounded, open subset of a Banach space X. Write

$$\Sigma_k(\bar{D}) = \{\phi : \bar{D} \to X; \phi = I - f, f \text{ a } k\text{-set contraction}\}$$

and
$$\Gamma(\bar{D}) = \{\phi : \bar{D} \to X; \phi = I - f, f \text{ condensing}\}.$$

In the terminology of Section 5.1, $\Sigma_k(\bar{D})$ is an admissible collection of mappings whatever the value of k, and so is $\Gamma(\bar{D})$. We shall define the degree of mappings in $\Sigma_k(\bar{D})$ with $k < 1$, and then of mappings in $\Gamma(\bar{D})$. As usual, Σ_k and Γ are given the topology of uniform convergence.

The members of Σ_k and Γ have the pleasant properties required in Section 5.2; Σ_k and Γ are convex, and we have the following.

Lemma 6.2.1 *If $\phi \in \Gamma(\bar{D})$, then ϕ is bounded and proper (and therefore closed).*

Proof It is immediate from the definition of a condensing map that ϕ is bounded. To show that ϕ is proper, consider a compact subset S of X, and let $S_1 = \phi^{-1}(S)$. If $\phi = I - f$, then $S_1 \subset f(S_1) + S$; we deduce (using Theorem 6.1.3(1) and Theorem 6.1.6) that $\alpha(S_1) \leqslant \alpha(f(S_1)) + \alpha(S)$. Since S is compact, $\alpha(S) = 0$; hence $\alpha(S_1) \leqslant \alpha(f(S_1))$. But $\alpha(f(S_1)) < \alpha(S_1)$ unless S_1 is compact, so we have a contradiction if S_1 is not compact.

Corollary 6.2.2 *If $\phi \in \Sigma_k(\bar{D})$ with $k < 1$, then ϕ is bounded and proper.*

Proof Merely observe that $\Sigma_k(\bar{D}) \subset \Gamma(\bar{D})$ for $k < 1$.

Define $\Delta_1 = \operatorname{co} f(\bar{D})$ and, inductively,

$$\Delta_n = \overline{\operatorname{co}}\, f(\Delta_{n-1} \cap \bar{D}) \quad (n = 2, 3, \ldots).$$

Also write $\Delta = \bigcap_{n=1}^{\infty} \Delta_n$. We note first that Δ is a closed, convex set. (Δ_n) is a nested sequence of sets: $\Delta_1 \supset \Delta_2 \supset \Delta_3 \ldots$ For certainly $\Delta_1 \supset \Delta_2$, and if we suppose that $\Delta_{k-1} \supset \Delta_k$, we find that

$$\Delta_k = \overline{\operatorname{co}}\, f(\Delta_{k-1} \cap \bar{D}) \supset \overline{\operatorname{co}}\, f(\Delta_k \cap \bar{D}) = \Delta_{k+1};$$

so, by induction, $\Delta_{n-1} \supset \Delta_n$ $(n = 1, 2, 3, \ldots)$. Lastly, $f(\Delta \cap \bar{D}) \subset \Delta$; this is proved by noting that $f(\Delta_n \cap \bar{D}) \subset \overline{\operatorname{co}}\, f(\Delta_n \cap \bar{D}) = \Delta_{n+1} \subset \Delta_n$ for all n.

Suppose that f is a k-set contraction with $k < 1$. Then

$$\alpha(\Delta_n) = \alpha(f(\Delta_{n-1} \cap \bar{D})) \leqslant \alpha(f(\Delta_{n-1})) \leqslant k\alpha(\Delta_{n-1}),$$

whence $\alpha(\Delta_n) \leqslant k^n \alpha(\bar{D})$. It follows that $\alpha(\Delta_n) \to 0$ as $n \to \infty$, so that, by Theorem 6.1.2, Δ is non-empty and compact. By Dugundji's extension of Tietze's theorem (Theorem 4.4.5) there is a mapping $\tilde{f}$ from $\bar{D}$ into Δ which coincides with f on $\bar{D} \cap \Delta$ (Δ is, of course, convex). Since the range of $\tilde{f}$ is compact, it is possible to speak of the (Leray–Schauder) degree of $I - \tilde{f}$ relative to D. If there is $x \in \bar{D}$

such that $\tilde{f}(x) = x$, then clearly $x \in \Delta$, and $\tilde{f}(x) = f(x)$. Therefore f and $\tilde{f}$ have the same fixed points, so that ϕ is unchanged on its zero set – the set that matters in this context. Also $d(I - \tilde{f}, D, 0)$ is defined provided $0 \notin (I - f)(\partial D) = \phi(\partial D)$. Let $\tilde{\tilde{f}} : \bar{D} \to \Delta$ be another continuous function which agrees with f on $\bar{D} \cap \Delta$, and consider the homotopy

$$h(t) = I - t\tilde{f} - (1 - t)\tilde{\tilde{f}} \quad (0 \leqslant t \leqslant 1).$$

Since $\tilde{f}$ and $\tilde{\tilde{f}}$ both map $\bar{D}$ into the convex set Δ, $h(t)(x) = 0$ only if $x \in \Delta$. But $\tilde{f} = \tilde{\tilde{f}} = f$ on Δ, so that $h(t)(x) = 0$ only if $f(x) = x$. We see then that $0 \notin h(t)(\partial D)$ for $0 \leqslant t \leqslant 1$. Therefore, by the homotopy invariance of Leray–Schauder degree, $d(I - \tilde{f}, D, 0) = d(I - \tilde{\tilde{f}}, D, 0)$. The following definition is consequently meaningful.

Definition 6.2.3 Let $\phi \in \Sigma_k(\bar{D})$ with $k < 1$. If $0 \notin \phi(\partial D)$, define the degree $d(\phi, D, 0)$ to be the Leray–Schauder degree $d(\tilde{\phi}, D, 0)$, where $\tilde{\phi} = I - \tilde{f}$ and $\tilde{f} : \bar{D} \to \Delta$ is any extension of $f|_{\bar{D} \cap \Delta}$ (as described above). For $p \notin \phi(\partial D)$, define $d(\phi, D, p)$ to be $d(\phi - p, D, 0)$, where $\phi - p$ denotes the mapping $x \mapsto \phi(x) - p$.

Lemma 6.2.4 *Suppose that S is a compact, convex set containing Δ such that f maps $\bar{D} \cap S$ into S. If $\hat{f} : \bar{D} \to S$ agrees with f on $\bar{D} \cap S$, then $d(\phi, D, 0) = d(I - \hat{f}, D, 0)$.*

Proof Suppose that $d(\phi, D, 0) = d(I - \tilde{f}, D, 0)$, where $\tilde{f}$ is chosen as in Definition 6.2.3. Define $h(t) = I - t\tilde{f} - (1 - t)\hat{f}$ $(0 \leqslant t \leqslant 1)$. If $x \in \partial D$ and $h(t)(x) = 0$, we must have $x \in S$, for $\tilde{f}(x) \in S$ and $\hat{f}(x) \in S$. Thus $x \in \bar{D} \cap S$, so $\hat{f}(x) = f(x)$; hence $x = t\tilde{f}(x) + (1 - t) f(x)$. Now $f(x) \in \Delta_1$, and $\tilde{f}(x) \in \Delta \subset \Delta_1$; hence $x \in \Delta_1$. Then $f(x) \in \Delta_2$; since also $\tilde{f}(x) \in \Delta_2$, we have $x \in \Delta_2$. Continuing inductively, we find that $x \in \Delta$. In that case, $\tilde{f}(x) = f(x)$, and $x = f(x)$. By the invariance under homotopy of the Leray–Schauder degree, it follows that, provided $0 \notin \phi(\partial D)$, $d(I - \hat{f}, D, 0) = d(I - \tilde{f}, D, 0)$, as required.

We must justify the use of the word 'degree' in Definition 6.2.3 by checking that the axioms given in Section 5.1 are satisfied.

Theorem 6.2.5 *Definition* 6.2.3 *does define a topological degree.*

Proof Note that axiom (IV) is incorporated in the definition.

As remarked after the statement of the axioms, (II) and (III) need only be verified when $p = 0$.

(1) Suppose that $p \in D; d(I, D, p) = d(I - p, D, 0)$. In this case f is the constant mapping $x \mapsto p, \Delta = \{p\}$, and we can take $\tilde{f} = f$. We know that $d(I - p, D, 0) = 1$ if d is the Leray–Schauder degree; it follows that $d(I, D, p) = 1$ in the present case also.

(2) Let D_1 and D_2 be disjoint open sets contained in D, with $0 \notin \phi(\bar{D} \backslash (D_1 \cup D_2))$. Choose a function $\tilde{f}$ in accordance with Definition 6.2.3 so that $d(\phi, D, 0) = d(I - \tilde{f}, D, 0)$. Just as Δ was defined in terms of D, so define Δ' and Δ'' in terms of D_1 and D_2, respectively; clearly $\Delta' \subset \Delta$ and $\Delta'' \subset \Delta$. By Lemma 6.2.4, $d(\phi, D_1, 0) = d(I - \tilde{f}, D_1, 0)$ and $d(\phi, D_2, 0) = d(I - \tilde{f}, D_2, 0)$. Axiom (II) follows now from the corresponding result for the Leray–Schauder degree:

$$d(I - \tilde{f}, D, 0) = d(I - \tilde{f}, D_1, 0) + d(I - \tilde{f}, D_2, 0).$$

(3) Suppose that $h: [0, 1] \to \Sigma_k(\bar{D})$ is continuous, with $0 \notin h(t)(\partial D)$ for $0 \leqslant t \leqslant 1$. Write $H(t, x) = x - h(t)(x)$ and $H(t)$ for the mapping $x \mapsto H(t, x)$. Because $H(t)$ is a k-set contraction on $\bar{D}$ for all $t \in [0, 1]$, H is a k-set contraction on $[0, 1] \times \bar{D}$. We can therefore define sets Δ_n^* $(n = 1, 2, \ldots)$:

$$\Delta_1^* = \overline{\text{co}}\, H([0, 1] \times \bar{D}); \quad \Delta_n^* = \overline{\text{co}}\, H([0, 1] \times \bar{D} \cap \Delta_{n-1}^*);$$

let $\Delta^* = \bigcap_{n=1}^{\infty} \Delta_n^*$. Then Δ^* is a convex, compact set and H maps $[0, 1] \times \bar{D} \cap \Delta^*$ into Δ^*. Choose a continuous mapping $\tilde{H}: [0, 1] \times \bar{D} \to \Delta^*$ which agrees with H on $([0, 1] \times \bar{D} \cap \Delta^*)$. If we denote the mapping $x \mapsto \tilde{H}(t, x)$ by $\tilde{H}(t)$, Lemma 6.2.4 enables us to say that $d(I - \tilde{H}(t), D, 0) = d(I - H(t), D, 0) = d(h(t), D, 0)$. Now $\tilde{H}(t)$ is a homotopy of compact mappings; moreover, if $x \in \partial D$ and $x = \tilde{H}(t)(x)$, then $x \in \Delta^* \cap \bar{D}$, whence $h(t)(x) = 0$, contrary to hypothesis. By the invariance under homotopy of the Leray–Schauder degree, $d(I - \tilde{H}(t), D, 0)$ is independent of $t \in [0, 1]$; hence $d(h(t), D, 0)$ is also independent of $t \in [0, 1]$.

We come now to the question of the uniqueness of the degree we have defined for $\Sigma_k(\bar{D})$. There is no question that it coincides with the Leray–Schauder degree when $k = 0$ – that is clear from the definition as well as having been proved in Theorem 5.3.1. We can show, moreover, that no other essentially different definition is possible for any $k < 1$ (Nussbaum has also proved this).

Theorem 6.2.6 *If $\phi \in \Sigma_k(\bar{D})$ $(k < 1)$ and the degree of ϕ at p relative to D is defined, then its value is necessarily given by Definition* 6.2.3.

Proof Suppose that d is a topological degree and that $d(\phi, D, 0)$ is defined (so that, in particular, $0 \notin \phi(\partial D)$). Let $\phi = I - f$ and let $\tilde{f}$ be as used in Definition 6.2.3. Consider

$$h(t) = I - tf - (1-t)\tilde{f} \quad (0 \leqslant t \leqslant 1).$$

Suppose that $h(t)(x) = 0$ with $x \in \partial D$. Now $\tilde{f}(x) \in \Delta \subset \Delta_1$ and $f(x) \in \Delta_1$, whence $x \in \Delta_1$; then $f(x) \in \Delta_2$, and $x \in \Delta_2$. Continuing in this way, $x \in \Delta$. But $f = \tilde{f}$ on Δ, so that $x = f(x)$. This is impossible, for we supposed that $0 \notin \phi(\partial D)$. Applying axiom (III) to the homotopy $h(t)$, we have that $d(\phi, D, 0) = d(I - \tilde{f}, D, 0)$. However $\tilde{f}$ is compact, so that $d(I - \tilde{f}, D, 0)$ is uniquely determined (Theorem 5.3.1). Hence $d(\phi, D, 0)$ is also uniquely determined. By axiom (IV), the same is true for $d(\phi, D, p)$.

Remark If f is a local k-set contraction, then $d(I - f, D, p)$ can be defined if $f^{-1}(p)$ is compact. For it can then be shown that f is a k-set contraction on an open set containing $f^{-1}(p)$.

We now proceed to define the degree of mappings belonging to $\Gamma(\bar{D})$ and $\Sigma_1(\bar{D})$; the two classes differ in that $\phi \in \Sigma_1(\bar{D})$ may not be closed. We shall define a degree for the whole of Σ_1, in contrast to Nussbaum (1972), who deals with those mappings in Σ_1 which are closed. Recall that $\Gamma \subset \Sigma_1$.

Suppose that $\phi = I - f \in \Sigma_1(\bar{D})$ and that $0 \notin \overline{\phi(\partial D)}$. Let $\eta = \rho(0, \overline{\phi(\partial D)})$; by hypothesis, $\eta > 0$. Choose a strict set contraction $g : \bar{D} \to X$ such that

$$\| f(x) - g(x) \| < \tfrac{1}{3}\eta \quad (x \in \bar{D});$$

such a choice is certainly possible – for example, $g = \lambda f$ will suffice if $\lambda < 1$ and $1 - \lambda$ is sufficiently small. Writing $\psi = I - g$, and noting that $\psi(\partial D)$ is closed, we see that $\rho(0, \psi(\partial D)) \geqslant \tfrac{2}{3}\eta$. Our aim is to justify defining $d(\phi, D, 0)$ to be $d(\psi, D, 0)$. Suppose that g_1 is another strict set contraction satisfying $\|f(x) - g_1(x)\| < \frac{1}{3}\eta$ for $x \in \bar{D}$. If g is a k_1-set contraction and g_1 a k_2-set contraction, then both g and g_1 are k-set contractions with $k = \max(k_1, k_2)$. Consider the homotopy

$$h(t) = I - tg - (1-t)g_1 \quad (0 \leqslant t \leqslant 1).$$

If $h(t)(x) = 0$ with $x \in \partial D$, then, because $\|g(x) - g_1(x)\| < \frac{2}{3}\eta$, we have $\|x - g_1(x)\| < \frac{2}{3}\eta$. However, g_1 was so chosen that $\|x - g_1(x)\| \geqslant \frac{2}{3}\eta$ for such x; therefore $0 \notin h(t)(\partial D)$ for $0 \leqslant t \leqslant 1$. We now use the invariance under homotopy of the degree of mappings in $\Sigma_k(\bar{D})$ to deduce that $d(\psi, D, 0) = d(\psi_1, D, 0)$, where $\psi_1 = I - g_1$.

Definition 6.2.7 Suppose that $\phi = I - f \in \Sigma_1(\bar{D})$ and $0 \notin \overline{\phi(\partial D)}$. Define $d(\phi, D, p)$ to be $d(I - g, D, 0)$, where g is any strict set contraction satisfying $\|f(x) - g(x)\| < \delta \, (x \in \bar{D})$, where δ is suitably small. If $p \notin \overline{\phi(\partial D)}$ define $d(\phi, D, p)$ to be $d(\phi - p, D, 0)$.

Remarks In the definition, we may take $\delta = \frac{1}{3}\rho(0, \overline{\phi(\partial D)})$. If $\phi \in \Gamma(\bar{D})$, then $d(\phi, D, p)$ is defined whenever $p \notin \phi(\partial D)$ (as opposed to $p \notin \overline{\phi(\partial D)}$).

Theorem 6.2.8 *Definition* 6.2.7 *does define a topological degree for* $\Sigma_1(\bar{D})$.

Proof Axioms (I) and (IV) need no detailed checking. As for axiom (II), let D_1 and D_2 be open subsets of D with $0 \notin \overline{\phi(\bar{D} \backslash (D_1 \cup D_2))}$. Writing $\phi = I - f$, choose a strict set contraction g such that $\|g(x) - f(x)\| < \delta$ for $x \in \bar{D}$ and $d(\phi, D, 0) = d(I - g, D, 0)$. Let $\psi = I - g$; if δ is sufficiently small, then $0 \notin \psi(\bar{D} \backslash (D_1 \cup D_2))$ (ψ is closed). Applying axiom (II) to the degree of perturbations of the identity by strict set contractions,

$$d(\psi, D, 0) = d(\psi, D_1, 0) + d(\psi, D_2, 0).$$

But, for $i = 1, 2, d(\psi, D_i, 0) = d(\phi, D_i, 0)$; thus axiom (II) is verified for $\Sigma_1(\bar{D})$.

To prove that axiom (III) is satisfied, we argue as in Theorem 5.1.6 and Corollary 5.1.8. Given $\phi = I - f \in \Sigma_1(\bar{D})$ with $0 \notin \overline{\phi(\partial D)}$, set

$$U(\phi, r) = \{\psi \in \Sigma_1(\bar{D});\ \|\psi(x) - \phi(x)\| < r \quad (x \in \bar{D})\},$$

where $r = \frac{1}{3}\rho(0, \overline{\phi(\partial D)})$. If $\psi = I - g \in U(\phi, r)$, then there exists $\lambda \in (0, 1)$ such that $I - \lambda f, I - \lambda g \in U(\phi, r)$ and $d(\phi, D, 0) = d(I - \lambda f, D, 0)$, $d(\psi, D, 0) = d(I - \lambda g, D, 0)$. The homotopy $H(t) = t(I - \lambda f) + (1 - t)(I - \lambda g)$ is such that, for $0 \leqslant t \leqslant 1$, (i) $H(t) \in U(\phi, r)$, whence $0 \notin \overline{H(t)(\partial D)}$, and (ii) $H(t) \in \Sigma_\lambda(\bar{D})$. Then, by axiom (III) for λ-set contractions, $d(I - \lambda f, D, 0) = d(I - \lambda g, D, 0)$; so $d(\phi, D, 0) = d(\psi, D, 0)$. Therefore $d(., D, 0)$ is a continuous mapping of $\Sigma_1(\bar{D})$ (with the topology of uniform convergence) into the integers. It follows (compare Corollary 5.1.8) that if $h : [0, 1] \to \Sigma_1(\bar{D})$ is a homotopy, then, provided $0 \notin \overline{h(t)(\partial D)}$ for $0 \leqslant t \leqslant 1$, $d(h(t), D, 0)$ is independent of $t \in [0, 1]$.

The next question that naturally arises is the following: is the degree $d(\phi, D, p)$ with $\phi \in \Sigma_1(\bar{D})$ uniquely determined?

Theorem 6.2.9 *For $\phi \in \Sigma_1(\bar{D})$, $d(\phi, D, p)$, if defined, is uniquely determined by Definition* 6.2.7.

Proof If d is a degree for the collection of mappings $\Sigma_1(\bar{D})$, then, by Theorem 5.1.6, $d(\phi, D, p) = d(\psi, D, p)$ when $\|\phi(x) - \psi(x)\|$ is uniformly small. But $d(\phi, D, p)$ was defined in terms of the degree of an approximation ψ of ϕ, itself uniquely determined (Theorem 6.2.6). Therefore $d(\phi, D, p)$ is uniquely determined.

In the next chapter, we discuss various concepts of a generalised degree; one such example is that of Browder and Nussbaum (1968). Nussbaum (1972) shows that if this degree is defined for a mapping $\phi = I - f$, where f happens to be a strict contraction, then it coincides with the degree of ϕ *qua* member of $\Sigma_1(\bar{D})$. Theorem 6.2.9 strengthens this conclusion, showing that any 'generalised degree' coincides with the degree of Definition 6.2.7 for any mapping for which both are defined. As a particular case, the degree introduced by Fenske (1971) for mappings $I - f$ with f a strict set contraction coincides with the degree defined here for such mappings.

By means similar to the above, the degree of $I - f$ can be defined when f is a 1-ball contraction. If it so happens that f is both a 1-set contraction and a 1-ball contraction, the two degrees coincide.

6.3 Fixed point theorems

Many authors have proved fixed point theorems of various kinds for condensing maps and k-set contractions. (The reader will find a very useful account of these theorems in Petryshyn (1973).) We shall discuss such results from the viewpoint of degree theory. As usual, the ambient space X is a real Banach space and D is an open, bounded subset. One of the earliest results in the field was that of Darbo (1955).

Theorem 6.3.1 *Let S be a closed, bounded, convex subset of X, and suppose that $f: S \to S$ is a strict set contraction. Then f has a fixed point.*

The same result was proved by Sadovskii (1967) for condensing maps. Both can be deduced from the general theorem given below; this is similar to theorems found in Nussbaum (1971*a*, 1972), Petryshyn (1971, 1973) and elsewhere.

Theorem 6.3.2 *Suppose that S is a bounded subset of X with non-empty interior D. Let $f: S \to X$ be a* 1*-set contraction such that* (i) $(I-f)(\bar{D})$, $(I-f)(\partial D)$ *are closed, and* (ii) *there is $w \in D$ such that*

$$f(x) - w \neq m(x-w) \quad (x \in \partial D, m > 1).$$

Then f has a fixed point in D.

Proof Without loss of generality, we take $w = 0$. We apply Theorem 5.1.10. Since $I - f \in \Sigma_1(\bar{D})$, we see that $I - tf \in \Sigma_t(\bar{D})$; hence $I - tf$ is a closed mapping for $0 \leqslant t < 1$. The conditions of Theorem 5.1.10 are all satisfied, whence the result.

Remarks on Theorem 6.3.2 (1) Condition (ii) of the theorem is certainly satisfied if D is convex and $f(\partial D) \subset D$. It is also satisfied if

$$\| f(x) - x \|^2 \geqslant \| f(x) - w \|^2 - \| x - w \|^2 \quad (x \in \partial D);$$

this is a condition used by Altman (1957).

(2) Condition (i) is satisfied *a fortiori* if f is condensing or a strict set contraction. Darbo's theorem follows from Theorem 6.3.2 by

extending $f: S \to S$ to a mapping $\tilde{f}$ of a ball of suitably large radius into S, and then applying the theorem to $\tilde{f}$.

(3) A special case of Theorem 6.3.2 is the well-known result of Krasnosel'skii (1955): if D is convex and $f: \bar{D} \to \bar{D}$ is of the form $g + h$, where g is a contraction mapping and h is compact, then f has a fixed point. We noted in Section 6.1 that a mapping of this form is a strict set contraction.

(4) The theorem is not valid if condition (i), or something similar, is missing. Let X be the Hilbert space l_2, and let

$$g(x) = (0, x_1, x_2, \ldots), \quad h(x) = (\sqrt{1 - \|x\|^2}, 0, 0, \ldots),$$

where $x = (x_1, x_2, \ldots)$. Then $f = g + h$ maps the closed unit ball $\bar{B}$ to itself. Now g is a 1-set contraction and h is compact; so f is a 1-set contraction. However, f has no fixed points in $\bar{B}$, for $f(\xi) = \xi$ implies that $\xi_1^2 = 1 - \|\xi\|^2$ and $\xi_1 = \xi_2 = \xi_3 = \ldots$; the latter can occur only if $\xi_i = 0$ $(i = 1, 2, 3, \ldots)$, and this contradicts the former.

(5) The theorem and its consequences apply to mappings which are 1-ball contractions; the proofs are entirely analogous to those given here for 1-set contractions.

Theorem 6.3.2 yields fixed point theorems for many classes of mappings which do not immediately appear to fall under its scope. A few of these will be mentioned.

A mapping $f: \bar{D} \to X$ *is non-expansive* if

$$\|f(x) - f(y)\| \leqslant \|x - y\| \quad (x, y \in \bar{D});$$

f is a *generalised contraction* if, for all $x \in \bar{D}$, there is $\alpha(x) < 1$ such that

$$\|f(x) - f(y)\| \leqslant \alpha(x)\|x - y\| \quad (y \in \bar{D}).$$

It is easy to see that $f = g + h$ is a 1-set contraction when h is compact and g is either non-expansive or a generalised contraction. Petryshyn (1973) shows that a generalised contraction satisfies the conditions of Theorem 6.3.2 when X is reflexive.

Nussbaum (1972) introduced a class of mappings which he called LANE (*locally almost non-expansive*). A continuous mapping $f: \bar{D} \to X$ is LANE if, for $x \in \bar{D}$ and $\varepsilon > 0$, there is a neighbourhood $U(x)$ of x in the weak topology such that

$$\| f(y) - f(z) \| \leqslant \| y - z \| + \varepsilon \quad (y, x \in U(x)).$$

Examples of LANE mappings are the semicontractive mappings introduced by Browder (1968) (f is *semicontractive* if there exists a mapping $V : \bar{D} \times \bar{D} \to X$ such that $f(x) = V(x, x)$ and (i) $V(.,y)$ is non-expansive, (ii) given z and ε, there is a weak neighbourhood $U(z)$ such that $\| V(x, y) - V(x, z) \| < \varepsilon$ for $x \in \bar{D}$ and $y \in U(z)$). Nussbaum showed that in a uniformly convex Banach space, a LANE mapping f is a 1-set contraction and $(I - f)(\bar{D})$ is closed, at least if D is convex. Under these conditions, therefore, f has a fixed point in D if it also satisfies (ii) of Theorem 6.3.2.

Browder (1968) also introduced mappings described as of semi-contractive type, strongly semicontractive type and weakly semi-contractive type. Without entering into the details, we simply say that these three related classes of mappings are all covered by the above fixed point theorem. The interested reader is referred to the paper of Petryshyn (1973) which was mentioned above.

We go on to prove another fixed point theorem, of rather a different kind to those mentioned up to now. It is due to Edmunds, Potter and Stuart (1972), and is a generalisation of a theorem of Krasnosel'skii (1964). It is worth noting that there is a general version in the spirit of Theorem 5.1.10.

Definition 6.3.3 The subset S of X is a *cone* if (a) S is non-empty and closed, (b) S contains $x + y$ and αx whenever it contains x and y, and $\alpha \geqslant 0$, (c) S contains both x and $-x$ only if $x = 0$.

Theorem 6.3.4 *Let S be a cone and $f : X \to S$ be a k-set contraction with $k < 1$. Suppose that there are numbers r and R, with $0 < r < R$, such that $x - f(x) \notin S$ if $x \in S$ and $\| x \| = r$, while $f(x) - x \notin S$ if $x \in S$ and $\| x \| = R$. Then f has a fixed point ξ in S such that $r < \| \xi \| < R$.*

Proof Let B_ρ denote the open ball with centre 0 and radius ρ, and let S_ρ be the corresponding sphere. We regard f as a mapping $\bar{B}_R \to S$, and use the degree theory developed for mappings in Σ_k, recalling that these are closed mappings when $k < 1$.

(i) Define $H(t) = I - tf$ for $0 \leqslant t \leqslant 1$; then H is a continuous mapping of $[0,1]$ into $\Sigma_k(\bar{B}_R)$. If $H(t)(x)=0$ for $x \in S_R$, then, since $f(x) \in S$ and S is a cone, we have $f(x)-x=(1-t)f(x) \in S$; this is contrary to our hypotheses. It follows by invariance under homotopy that $d(I-f, B_R, 0)=d(I, B_R, 0)$; therefore we have $d(I-f, B_R, 0)=1$.

(ii) For $x \in S_r$ we have that $x-f(x) \notin S$ – for otherwise $x=(x-f(x))+f(x) \in S$ (since S is a cone), and we would have $x \in S$, $\|x\|=r$ and $x-f(x) \in S$, which we have supposed not to happen. Hence $d(I-f, B_r, 0)$ is defined.

(iii) Choose any $p \in S \backslash \{0\}$; there is a positive integer n such that $w-p \notin S$ if $\|w\| \leqslant r/n$. Define

$$h(t)=I-f-tnp \quad (0 \leqslant t \leqslant 1).$$

Again h is a continuous mapping of $[0,1]$ into $\Sigma_k(\bar{B}_r)$. If $h(t)(x)=0$ for $x \in S_r$, we have $x-f(x)=tnp \in S$, a contradiction. Hence $d(I-f, B_r, 0)=d(I-f-np, B_r, 0)$.

(iv) We next show that $np \notin (I-f)(\bar{B}_r)$. If $x \in \bar{B}_r$ and $x-f(x)=np$, then $x-np \in S$, whence $x/n-p \in S$. Since $\|x/n\| \leqslant r/n$, this is impossible, by the choice of n. Consequently $np \notin (I-f)(\bar{B}_r)$, and $d(I-f-np, B_r, 0)=0$ (this by Theorem 5.1.4).

(v) Let $D=B_R \backslash \bar{B}_r$. By axiom (II) and (i), (iv) above,

$$d(I-f, D, 0)=d(I-f, B_R, 0)-d(I-f, B_r, 0)=1.$$

It follows that f has a fixed point in D.

As well as to derive fixed point theorems, degree theory can be used to prove mapping theorems. For example, Webb (1971) gives some results of this kind, including one on the invariance of domain: if $D \subset X$ is open, $f: \bar{D} \to X$ is condensing and $I-f$ is one to one, then $(I-f)(D)$ is open. The proof depends on the fact that the degree of an odd mapping is odd.

To end this chapter we mention one application of some of the fixed point theorems we have proved. Consider the differential equation

$$\frac{dx}{dt}=f(x), \tag{6.3.1}$$

where $x \in X$ (a Banach space) and $t \in \mathbf{R}$. When X is infinite dimensional the question of the existence of solutions of (6.3.1) is much more complicated than it is when X is finite dimensional. Suppose that U is an open subset of X with $x_0 \in U$, and let $f: U \to X$ be continuous. Suppose also that there are constants M and b such that $\| f(x) \| \leqslant M$ when $x \in Q = \{x; \| x - x_0 \| \leqslant b\}$. Define

$$L(\varepsilon) = \sup\left\{\frac{\alpha(f(S))}{\alpha(S)}; S \subset Q, \alpha(S) \geqslant \varepsilon\right\}.$$

It can be shown that, if

$$\int_0^1 \frac{d\varepsilon}{\varepsilon L(\varepsilon)} = \infty,$$

then there is $a > 0$ and a function $\phi(t)$, defined and (Fréchet) differentiable for $0 \leqslant t < a$ such that

$$\frac{d\phi}{dt} = f(\phi) \quad (0 \leqslant t < a)$$

and $$\phi(0) = x_0.$$

Details of this and related results may be found in, for example, Cellina (1971).

7
Generalised degree

We continue the process of extending the collection of mappings for which it is possible to define a degree. Several of the developments related in this chapter were inspired by Browder and his collaborators. In contrast with earlier chapters all the details will not be given, rather we give a resumé with references. The chapter falls into three distinct parts: in the first we look at mappings of the form 'homeomorphism + compact mapping', and generalisations of them; in the second part a multivalued degree is defined for perturbations of the identity by A-proper maps; finally, mappings which are themselves multivalued are considered.

7.1 Intertwined representations

We shall in this section consider mappings such as those studied by Browder (1968), Browder and Nussbaum (1968), and Browder and Gupta (1969). We are more restrictive in our hypotheses than is Browder; the way to generalise is, however, fairly clear.

Definition 7.1.1 Suppose that X is a Banach space and D a bounded, open subset of X. A continuous mapping $h:\bar{D} \to X$ is a *permissible homeomorphism* of $\bar{D}$ if (i) $h(D)$ is bounded, (ii) h maps D homeomorphically onto $h(D)$, and (iii) h maps $\bar{D}$ homeomorphically onto $\overline{h(D)}$.

Let $P(\bar{D})$ be the set of all permissible homeomorphisms of $\bar{D}$; as before, let $K(\bar{D})$ be the set of compact mappings of $\bar{D}$ into X. Denote by β the collection of bounded, open subsets of X. For each $D \in \beta$, choose a convex subset $P_0(\bar{D})$ of $P(\bar{D})$ such that (*a*) $I \in P_0(\bar{D})$, (*b*) $P_0(\bar{D})$ is closed under the addition of constants, and (*c*) if $D \subset D_1 \in \beta$ and $h \in P_0(\bar{D}_1)$, then $h|_{\bar{D}} \in P_0(\bar{D})$. Now write $P_1(\bar{D})$

for the set $\{f = h + C; h \in P_0(\bar{D}), C \in K(\bar{D})\}$; then, in the notation of Chapter 5,

$$P_1(\beta) = \{P_1(\bar{D}); D \in \beta\}$$

is an admissible class of mappings. Naturally, both $P_0(\bar{D})$ and $P_1(\bar{D})$ are considered to have the topology of uniform convergence.

Definition 7.1.2 Let $f = h + C \in P_1(\bar{D})$, and suppose that $p \notin f(\partial D)$. Define $d(f, D, p)$ to be the Leray–Schauder degree $d(I + Ch^{-1}, h(D), p)$.

This definition is meaningful, for $h(D)$ is open and bounded, $Ch^{-1} \in K(h(D))$, and $p \notin (I + Ch^{-1})(\partial h(D))$. Moreover, Browder shows that the definition is independent of the particular representation of f as a sum $h + C$ with $h \in P_0(\bar{D})$ and $C \in K(\bar{D})$; he also proves that the function satisfies the axioms set out in Chapter 5, so justifying its being called a topological degree. However, the degree of f is defined only relative to the chosen class $P_1(\beta)$, which depends in turn on $P_0(\beta)$; this introduces a certain ambiguity.

More generally we consider mappings defined as follows.

Definition 7.1.3 The continuous mapping $\phi: \bar{D} \to X$ has an *intertwined representation* with respect to $P_0(\bar{D})$ if there is a continuous mapping $S: \bar{D} \times \bar{D} \to X$ such that $\phi(x) = S(x, x)$, and S satisfies the following two conditions:

(1) For each $v \in \bar{D}$, the mapping $S_v : x \mapsto S(x, v)$ of $\bar{D}$ into X belongs to $P_0(\bar{D})$.

(2) The mapping $S^* : v \mapsto S_v$ of $\bar{D}$ into $P_0(\bar{D})$ is compact.

We say that S is a *representation* of ϕ in $P_0(\bar{D})$.

For $D \in \beta$, let $P_2(\bar{D})$ denote the collection of mappings $\bar{D} \to X$ which have an intertwined representation with respect to $P_0(\bar{D})$. It is easily seen that $P_1(\bar{D}) \subset P_2(\bar{D})$. The members of $P_2(\bar{D})$ are proper, bounded mappings, and the class $P_2(\beta) = \{P_2(\bar{D}); D \in \beta\}$ is admissible.

Given $\phi \in P_2(\bar{D})$ we wish to define the degree of ϕ. Again this will be in the context of the chosen sets $P_0(\bar{D})$. The definition requires some justification.

Definition 7.1.4 Suppose that $\phi \in P_2(\bar{D})$ and $0 \notin \phi(\partial D)$. Let $D_0 = \{v \in D; 0 \in S_v(D)\}$ and define $C: \bar{D}_0 \to \bar{D}$ by $C(v) = S_v^{-1}(0)$. Then $d(\phi, D, 0)$ is defined to be the Leray–Schauder degree $d(I - C, D_0, 0)$; $d(\phi, D, p)$ is defined to be $d(\phi - p, D, 0)$ if $p \neq 0$.

That Definition 7.1.4 is sensible follows from the following lemma, whose proof we omit.

Lemma 7.1.5 *(1) D_0 is open. (2) $\bar{D}_0 \subset \{v \in \bar{D}; 0 \in S_v(\bar{D})\}$. (3) C is a compact mapping.*

Since D_0 is obviously bounded, $d(I - C, D_0, 0)$ is certainly defined if $0 \notin (I - C)(\partial D_0)$. Suppose, then, that $C(x) = x$ with $x \in \partial D_0$. We have $S(x, x) = \phi(x) = 0$. By Lemma 7.1.5(2), $x \in \partial D_0$ implies that $0 \in S_x(\partial D)$. It follows that $S_x^{-1}(0) \in \partial D$ – that is, $x = C(x) \in \partial D$. Hence $d(I - C, D_0, 0)$ is defined when $0 \notin \phi(\partial D)$.

Strictly, what has been defined is the degree of ϕ given a particular representation S; we should therefore write $d([\phi, S], D, 0)$, say, in place of $d(\phi, D, 0)$ to highlight this dependence on S. However, since $P_0(\bar{D})$ is convex, it turns out that $d([\phi, S], D, 0)$ is independent of the particular representation. This is deduced by Browder (1968) from a form of homotopy invariance which in turn depends on a result similar to Lemma 7.1.5.

There is comparatively little difficulty in showing that the axioms of Section 5.1 are satisfied by the function d. It is clear that $d(I, D, p) = 1$ if $p \in D$, simply by recalling the corresponding property for the Leray–Schauder degree. Homotopy invariance of d also follows from the properties of Leray–Schauder degree. As for the axiom of additivity, suppose that D_1 and D_2 are disjoint open sets contained in D, and that $0 \notin \phi(\bar{D} \backslash (D_1 \cup D_2))$. For D_1 and D_2, let $D_{1,0}$ and $D_{2,0}$, respectively, be the sets that are formed analogously to D_0. Then $(I - C)(x) \neq 0$ if $x \notin D_{1,0} \cup D_{2,0}$ (for $C(x) = x$ implies $S_x^{-1}(0) = x$, whence $S(x, x) = \phi(x) = 0$). Since $D_{1,0}$ and $D_{2,0}$ are disjoint, axiom (II) follows again from the corresponding property of the Leray–Schauder degree.

Theorem 7.1.6 *Definition* 7.1.4 *does define a topological degree* *for $P_2(\beta)$. Also the degree of ϕ is independent of the representation of ϕ in $P_0(\bar{D})$.*

Finally, consider mappings ϕ in the closure of $P_2(\bar{D})$ in $C(\bar{D})$, the space of continuous mappings $\bar{D} \to X$ with the topology of uniform convergence. Suppose that $p \notin \phi(\partial D)$. Choose a sequence ϕ_n in $P_2(\bar{D})$ such that $\phi_n \to \phi$. It is not difficult to prove that $d(\phi_n, D, p)$ tends to a limit as $n \to \infty$; $d(\phi, D, p)$ is defined to be this limit. It can also be shown that this does define a degree for mappings in $\overline{P_2(\bar{D})}$; we refer the reader to Browder's paper for details.

Remark We have throughout adopted the normalising axiom in the form '$d(I, D, p) = 1$ if $p \in D$'. By specifying the degree of the identity in this way, we are restricted in our choice of the set $P_0(\bar{D})$. A different set of possibilities for $P_0(\bar{D})$ is available to us if we choose to specify the degree of some other mapping, thereby inserting a new axiom of normalisation into the theory. This is perfectly possible, and is at the root of the difference between this account and that of Browder.

7.2 *A*-proper mappings

The definition and calculation of the degree of a compact perturbation of the identity $(I - f)$ depends on finding a uniform approximation to f whose range is finite dimensional. This technique can be used in more general circumstances, and is exploited by Browder and Petryshyn (1968, 1969) in their consideration of A-proper mappings (defined below). They defined a concept of degree whose most striking feature is that it is not single-valued; its values are subsets of the extended set of integers $\mathbf{Z} \cup \{-\infty, +\infty\}$. This is therefore not a topological degree in the sense of Chapter 5; we call it a *generalised degree*. We commence with some definitions.

Definition 7.2.1 Let X and Y be real Banach spaces. An *approximation scheme* Γ for mappings from X to Y consists of (*a*) two sequences $\{X_n\}$ and $\{Y_n\}$ of oriented finite dimensional spaces with X_n and Y_n of the same dimension (for each n), and (*b*) two sequences

$\{P_n\}$ and $\{Q_n\}$ of continuous mappings, where P_n maps X_n into X and Q_n maps Y into Y_n.

Definition 7.2.2 Let D be an open subset of X and ϕ a continuous mapping of $\bar{D}$ into Y. We say that ϕ is *A-proper* with respect to the approximation scheme Γ if, given a sequence (x_{n_j}) such that $x_{n_j} \in X_{n_j}, P_{n_j}(x_{n_j}) \in \bar{D}$ and $\| Q_{n_j}\phi P_{n_j}(x_{n_j}) - Q_{n_j}(y) \| \to 0$ for some $y \in Y$, there exists a subsequence $(x_{n_{j(k)}})$ and $x \in X$ such that $P_{n_{j(k)}}(x_{n_{j(k)}}) \to x$ and $\phi(x) = y$.

Notation It is convenient not to distinguish notationally between the norms of the various spaces that occur. In this section we regard X, Y and the scheme Γ as fixed; D will be an open subset of X, and the collection of A-proper mappings of $\bar{D}$ into Y will be denoted by $\mathscr{A}(\bar{D})$. We write $\phi_n = Q_n \circ \phi \circ P_n$ and $D_n = P_n^{-1}(D)$. Though D may be unbounded, we do require D_n to be bounded for each n.

Definition 7.2.3 Suppose that $\phi \in \mathscr{A}(\bar{D})$ and $p \notin \phi(\partial D)$. Define the *generalised degree* $\mathrm{Deg}(\phi, D, p)$ to be the following subset of $\mathbf{Z}' = \mathbf{Z} \cup \{-\infty, +\infty\}$:

$$\{\gamma \in \mathbf{Z}' ; \text{ there is a sequence } (n_j) \text{ such that } d(\phi_{n_j}, D_{n_j}, Q_{n_j}(p)) \to \gamma \text{ as } j \to \infty\}.$$

The degree $d_{n_j} = d(\phi_{n_j}, D_{n_j}, Q_{n_j}(p))$ is the degree defined for mappings between oriented finite dimensional spaces of the same dimension (see Section 1.4, Corollary 1.4.5). That the set $\mathrm{Deg}(\phi, D, p)$ is non-empty follows if d_{n_j} is defined for sufficiently large j. Certainly D_{n_j} is an open, bounded subset of X_{n_j} and $\phi_{n_j}: \bar{D}_{n_j} \to Y_{n_j}$ is continuous; by the A-properness of ϕ and the hypothesis $p \notin \phi(\partial D)$ it is shown quite easily that $Q_{n_j}(p) \notin \phi_{n_j}(\partial D_{n_j})$ if j is sufficiently large, so that d_{n_j} is defined.

Some of the main properties of the generalised degree are given in the next theorem. For homotopy invariance the sequence (Q_n) is assumed to be equi-uniformly continuous on bounded subsets of Y. This means that, given $\varepsilon > 0$ and a bounded subset B of Y, there is $\delta > 0$ such that $\| Q_n(y) - Q_n(x) \| < \varepsilon$ for all n and $x, y \in B$ with $\| x - y \| < \delta$.

Theorem 7.2.4 *Suppose that* $\phi \in \mathscr{A}(\bar{D})$ *and* $p \notin \phi(\partial D)$.

(1) *If* $\mathrm{Deg}(\phi, D, p) \neq \{0\}$, *there is* $x \in D$ *such that* $\phi(x) = p$.

(2) *Let* $h_t : [0, 1] \to \mathscr{A}(\bar{D})$ *be continuous* ($\mathscr{A}(\bar{D})$ *having the topology of uniform convergence). If* (Q_n) *is equi-uniformly continuous on bounded subsets of* Y *and* $p \notin h_t(\partial D)$ $(0 \leqslant t \leqslant 1)$, *then* $\mathrm{Deg}(h_t, D, p)$ *is independent of* $t \in [0, 1]$.

(3) *If* $D^{(1)}$ *is an open subset of* D *and* $p \notin \phi(\bar{D} \backslash D^{(1)})$, *then* $\mathrm{Deg}(\phi, D, p) = \mathrm{Deg}(\phi, D^{(1)}, p)$.

(4) *If* $D = D^{(1)} \cup D^{(2)}$, *where* $D^{(1)}$ *and* $D^{(2)}$ *are open, and* $p \notin \phi(\partial D^{(1)} \cup \partial D^{(2)} \cup (D^{(1)} \cap D^{(2)}))$ *then*

$$\mathrm{Deg}(\phi, D, p) \subset \mathrm{Deg}(\phi, D^{(1)}, p) + \mathrm{Deg}(\phi, D^{(2)}, p)$$

$$= \{\gamma ; \gamma = \gamma_1 + \gamma_2, \gamma_i \in \mathrm{Deg}(\phi, D^{(i)}, p) \quad (i = 1, 2)\}.$$

Equality holds if $\mathrm{Deg}(\phi, D^{(1)}, p)$ *or* $\mathrm{Deg}(\phi, D^{(2)}, p)$ *is a singleton.*

Remarks In (4) we adopt the convention that $(+\infty) + (-\infty) = \gamma$ for every $\gamma \in \mathbf{Z}'$. Parts (3) and (4) of the theorem correspond to the axiom of additivity for a single-valued topological degree.

Proof We shall illustrate the methods used by proving (1) and (3); the proofs of (2) and (4) are longer, and for these the reader is referred to the paper of Browder and Petryshyn (1969).

If $\mathrm{Deg}(\phi, D, p) \neq \{0\}$, then there is a sequence (n_j) such that $d(\phi_{n_j}, D_{n_j}, Q_{n_j}(p)) \neq 0$. Hence there is a sequence (x_{n_j}) with $x_{n_j} \in D_{n_j}$ such that $\phi_{n_j}(x_{n_j}) = Q_{n_j}(p)$. Since ϕ is A-proper, there is an x_0 and a subsequence $(x_{n_{j(k)}})$ of (x_{n_j}) such that $P_{n_j}(x_{n_j}) \to x_0 \in \bar{D}$ and $\phi(x_0) = p$. We are assuming that $p \notin \phi(\partial D)$; therefore $x_0 \in D$, as required.

As for (3), let $D_n^{(1)} = P_n^{-1}(D^{(1)})$. If, for infinitely many n, there exists $x_n \in D_n \backslash D_n^{(1)}$ with $\phi_n(x_n) = Q_n(p)$, then there is a sequence (y_{n_j}) with $y_{n_j} \in D \backslash D^{(1)}$ and $Q_{n_j}\phi(y_{n_j}) = Q_{n_j}(p)$. By the A-properness of ϕ, there are x_0 and a further subsequence $(x_{n_{j(k)}})$ such that $P_{n_{j(k)}}(x_{n_{j(k)}}) \to x_0 \in \bar{D}$ and $\phi(x_0) = p$. But $P_{n_{j(k)}}(x_{n_{j(k)}}) \in D \backslash D^{(1)}$, whence $p \in \phi(\bar{D} \backslash D^{(1)})$, contrary to hypothesis. We deduce that $Q_n(p) \notin \phi_n(D_n \backslash D_n{}^{(1)})$; so, since $Q_n(p) \notin \phi_n(\partial D_n)$ for large enough n, we have that $Q_n(p) \notin \phi_n(\bar{D}_n \backslash D_n^{(1)})$, whence $d(\phi_n, D_n, Q_n(p)) = d(\phi_n, D_n^{(1)}, Q_n(p))$ for sufficiently large n. From this it is clear that $\mathrm{Deg}(\phi, D, p) = \mathrm{Deg}(\phi, D^{(1)}, p)$.

If for ϕ, both $\mathrm{Deg}(\phi, D, p)$ and a single-valued degree $d(\phi, D, p)$ are defined, the relationship between the two is of obvious interest. The first requirement is to make certain that in such a case $\mathrm{Deg}(\phi, D, p)$ is a singleton (Browder and Petryshyn (1969) investigate this question). Suppose that $\mathrm{Deg}(\phi, D, p)$ is single-valued for ϕ belonging to some subclass $\mathscr{C}$ of $\mathscr{A}(\bar{D})$. Theorem 7.2.4 tells us that $\mathrm{Deg}(\phi, D, p)$ is then a topological degree for the mappings of $\mathscr{C}$. If we also know that the single-valued degree is uniquely defined for the members of $\mathscr{C}$ (as we do if $\mathscr{C}$ is covered by one of the cases studied in Chapter 6), then $\mathrm{Deg}(\phi, D, p) = \{d(\phi, D, p)\}$.

We turn now to a particular approximation scheme which is frequently encountered. Indeed, A-properness is often discussed entirely within this framework. We take $Y = X, P_n = I$ and Q_n a projection (that is, a continuous, linear mapping satisfying $Q_n^2 = Q_n$) for all n; we now require D to be bounded.

Definition 7.2.5 The real Banach space X is *projectionally complete* if (i) there is a sequence (X_n) of finite dimensional subspaces of X, (ii) there is a sequence (Q_n) of projections of X to X_n, and (iii) for all $x, Q_n(x) \to x$ as $n \to \infty$.

By the principle of uniform boundedness, there is $k \geqslant 1$ such that $\| Q_n \| \leqslant k$ for all n. In that case, X is said to be a π_k space.

When X is a π_k space, Wong (1971) defined the degree of A-proper mappings somewhat differently to the above. He used some ideas from non-standard analysis, and by so doing obtained equality in the additivity property (Theorem 7.2.4(4)).

When X is taken to be a π_1 space, Webb (1971) showed that $I - T$ is A-proper when T is ball-condensing. He also showed that the degree of $I - T$ as defined in Chapter 6 of this book is the same as its generalised degree *qua* A-proper mapping.

We introduce some more classes of mappings and then prove a fixed point theorem for some of them using the generalised degree we have defined for A-proper mappings.

Definition 7.2.6 Suppose that X is projectionally complete and that D is a bounded, open subset of X. The mapping $\phi : \bar{D} \to X$ is

P_*-compact if $\phi + \lambda I$ is A-proper for all $\lambda \geqslant 0$; if $\gamma > 0$, ϕ is *P_γ-compact* in case $\phi - \lambda I$ is A-proper for all $\lambda \geqslant \gamma$.

P_γ-compact mappings were investigated by Petryshyn and Tucker (1969) and Deimling (1970), amongst others. Related mappings were introduced by de Figueiredo (1967). P-compact (standing for 'projectionally compact') mappings are also defined in the literature (Petryshyn (1966) and elsewhere): ϕ is P-compact if $\phi - \lambda I$ is A-proper for all $\lambda > 0$. All these classes of mappings arise in the study of functional equations – and, in particular, partial differential equations; they encompass, for example, compact mappings and many kinds of mappings having a monotonicity or accretiveness type property.

In the proof of the following theorem, we make use of the facts that the identity mapping I is A-proper and that $k\phi$ is A-proper if $k \neq 0$ and ϕ is A-proper.

Theorem 7.2.7 *Let $\phi : \bar{D} \to X$ be P_*-compact. Suppose that $0 \in D$, $\phi(\bar{D})$ is bounded, and*

$$\phi(x) + \lambda x \neq 0 \quad (x \in \partial D, \lambda \geqslant 0).$$

Then there is $\xi \in D$ such that $\phi(\xi) = 0$.

Proof Consider the homotopy $h_t = (1-t)\phi + tI$ $(0 \leqslant t \leqslant 1)$. Since we may write $h_t = (1-t)[\phi + \tau I]$, where $\tau = t/(1-t)$, we see that $h_t \in \mathscr{A}(\bar{D})$ for $t \neq 1$; also $h_1 = I \in \mathscr{A}(\bar{D})$. Now $\phi(\bar{D})$ is bounded, so that $t \mapsto h_t$ is a continuous mapping of $[0,1]$ into $\mathscr{A}(\bar{D})$. Clearly $0 \notin h_1(\partial D)$ (for $0 \in D$); if $h_t(x) = 0$ with $x \in \partial D$ and $0 \leqslant t < 1$, then $\phi(x) + \lambda x = 0$, where $\lambda = t/(1-t) \geqslant 0$ – contradicting our hypothesis. The conditions of Theorem 7.2.4(2) are satisfied; therefore $\mathrm{Deg}(\phi, D, 0) = \mathrm{Deg}(I, D, 0)$. Now it can be seen that $\mathrm{Deg}(I, D, 0) = \{1\}$; thus, by Theorem 7.2.4(1), there is $\xi \in D$ satisfying $\phi(\xi) = 0$.

Corollary 7.2.8 *Let $\psi : \bar{D} \to X$ be P_1-compact. Suppose that $0 \in D$, $\psi(\bar{D})$ is bounded, and*

$$\psi(x) \neq \lambda x \quad (x \in \partial D, \lambda \geqslant 1).$$

Then ψ has a fixed point in D.

Proof Let $\phi = I - \psi$. Since ψ is P_1-compact, $\psi - \lambda I$ is A-proper for $\lambda \geqslant 1$; hence $\phi + \mu I$ is A-proper for $\mu \geqslant 0$ – that is, ϕ is P_*-compact. Also $\phi(x) + \mu x \neq 0$ for $x \in \partial D$ and $\mu \geqslant 0$. The existence of a fixed point of ϕ now follows from Theorem 7.2.7.

Note Another generalised degree was defined by Mawhin (1972). This he calls 'coincidence degree' and is for pairs of mappings (L, N), where L is linear but not necessarily invertible. The paper of Mawhin contains a complete account.

7.3 Multivalued mappings

Several instances are known of the extension of theorems concerning single-valued mappings to multivalued mappings (a multivalued mapping defined on a subset S of the space X is, of course, a mapping from S to 2^X, the power set of X). Such results have been studied by, for example, Eilenberg and Montgomery (1946), Granas and Jaworowski (1959) and Granas (1959*a*). Several fixed point theorems for multivalued mappings are given by Reich (1972). Some of the recent developments in degree theory have been concerned with defining a degree for certain multivalued mappings. Granas (1959*b*) introduced the degree of compact multivalued mappings *via* the methods of algebraic topology. Hukuhara (1967), Cellina and Lasota (1969) and Ma (1972) have all investigated the idea from an analytic viewpoint; we shall follow the method used by Cellina and Lasota. More general mappings (called 'ultimately compact') have been considered by Petryshyn and Fitzpatrick (1974, 1975). Some of these authors worked with locally convex topological vector spaces – we, however, persist with the setting of a Banach space (X).

The appropriate analogue of continuity when dealing with multivalued mappings is upper semicontinuity.

Definition 7.3.1 Let S be a subset of X. A mapping $\phi: S \to 2^X$ is *upper semicontinuous* if, given $x \in S$ and an open set V containing $\phi(x)$, there is an open set U containing x such that $\phi(U \cap S) \subset V$.

Definition 7.3.2 A mapping $\phi : S \to 2^X$ is *compact* if it is upper semicontinuous and maps bounded sets to relatively compact sets.

Let $\Gamma(X)$ be the collection of closed, convex subsets of X; suppose that D is an open, bounded subset of X. We shall define the degree of mappings from $\bar{D}$ to $\Gamma(X)$ which are of the form $I - f$ with f compact; we write $\kappa(\bar{D})$ for the collection of such mappings. For $\phi = I - f \in \kappa(\bar{D})$ we shall find single-valued, finite dimensional mappings to approximate f; the degree of ϕ will then be defined in terms of the degrees of these approximants.

Theorem 7.3.3 *Let S be a closed subset of X and let $T : S \to \Gamma(X)$ be upper semicontinuous. Given $\varepsilon > 0$, there is a continuous single-valued mapping T_ε defined on S such that if $x \in S$, there are $y \in S$ and $z \in T(y)$ with $\| y - x \| < \varepsilon$ and $\| z - T_\varepsilon(x) \| < \varepsilon$. Moreover, if the range of T is $R(T)$ and that of T_ε is $R(T_\varepsilon)$, then T_ε can be so chosen that $R(T_\varepsilon) \subset \overline{\text{co}}\, R(T)$; in particular, T_ε is compact when T is compact.*

Remarks The theorem states that, given a point ξ of the graph of T_ε, there is a point of the graph of T within ε of ξ. This observation leads to the following ideas. Let $\mathscr{G}(T_1), \mathscr{G}(T_2)$ be the graphs of two multivalued mappings T_1, T_2 of S into $\Gamma(X)$. Define the separation

$$\eta(T_1 ; T_2) = \sup_{\xi \in \mathscr{G}(T_1)} \rho^*(\xi, \mathscr{G}(T_2)),$$

where ρ^* is the metric on $X \times X$ given by

$$\rho^*((x, y), (u, v)) = \max(\| x - u \|, \| y - v \|).$$

The conclusion of Theorem 7.3.3 can be rephrased as: $\eta(T_\varepsilon ; T) < \varepsilon$. If (T_n) is a sequence of multivalued mappings such that $\eta(T_n ; T) \to 0$ as $n \to \infty$, we shall write $T_n \twoheadrightarrow T$.

Proof of Theorem 7.3.3 We shall need the idea of a 'star-refinement' of a covering of a set. Suppose that $\mathscr{U}$, $\mathscr{V}$ are coverings of a set A; $\mathscr{U}$ is a *star-refinement* of $\mathscr{V}$ if, given $W \in \mathscr{U}$, there is $V \in \mathscr{V}$ such that V contains the union of all $U \in \mathscr{U}$ which meet W. We use the fact that every open covering of S has an open, locally finite, star-refinement. (This, of course, holds more generally.) For further details, the reader may consult, for example, Chapter 6 of Willard (1970).

Because T is upper semicontinuous, given $\varepsilon > 0$ and $x \in S$, there is $\delta(x)$ with $0 < \delta(x) < \varepsilon$ such that $T(y) \subset B(T(x), \varepsilon)$ if $\| x - y \| < \delta(x)$ and $y \in S$. The collection $\mathscr{B} = \{B(x, \delta(x))\}$ is an open cover of S. Let $\mathscr{B}_1 = \{U_\alpha\}$ be a locally finite, star-refinement of $\mathscr{B}$, and let $\{f_\alpha\}$ be a continuous partition of unity subordinate to $\mathscr{B}_1$. Now select $z_\alpha \in T(S \cap U_\alpha)$, and define

$$T_\varepsilon(x) = \sum_\alpha f_\alpha(x) z_\alpha \quad (x \in S). \tag{7.3.1}$$

Take $x_0 \in S$; there is $y_0 \in S$ such that, if $x_0 \in U_\alpha \in \mathscr{B}_1$, then $U_\alpha \subset B(y_0, \delta(y_0))$ (for $B(y_0, \delta(y_0))$ certainly contains all the members of $\mathscr{B}_1$ which contain x_0). Then $\| x_0 - y_0 \| < \varepsilon$. Now $f_\alpha(x_0) \neq 0$ only if $x_0 \in U_\alpha$; all such U_α, we have seen, are contained in $B(y_0, \delta(y_0))$. So, for such α, $z_\alpha \in B(T(y_0), \varepsilon)$. By convexity it follows that $T_\varepsilon(x_0) \in B(T(y_0), \varepsilon)$. We can therefore take $z_0 \in T(y_0)$ such that $\| T_\varepsilon(x_0) - z_0 \| < \varepsilon$. This is as required. In addition, we have $R(T_\varepsilon) \subset \overline{\text{co}}(R(T))$, by (7.3.1). Since the space X is complete, $\overline{\text{co}}(R(T))$ is compact if $\overline{R(T)}$ is compact, and so $\overline{R(T_\varepsilon)}$ is compact if $\overline{R(T)}$ is compact.

Cellina (1969) used a result similar to Theorem 7.3.3 to prove Kakutani's fixed point theorem, that if S is compact and convex, and ϕ is an upper semicontinuous, multivalued mapping from S to the compact, convex subsets of S, then ϕ has a 'fixed point' in the sense that there is $\xi \in S$ with $\xi \in \phi(\xi)$. Cellina and Lasota (1969), using similar arguments, deduced the 'antipodal theorem': if $\phi = I - f \in \kappa(\bar{B})$, where B is the unit ball in X, and

$$\phi(x) \cap \lambda\phi(-x) = \emptyset \quad (0 \leqslant \lambda \leqslant 1, x \in \partial B),$$

then there is $\xi \in \bar{B}$ with $\xi \in f(\xi)$.

We now come to the promised definition of degree.

Definition 7.3.4 Suppose that $\phi = I - f \in \kappa(\bar{D})$ and $p \notin \phi(\partial D)$. Choose a sequence (f_n) of single-valued, compact mappings defined on $\bar{D}$ such that $f_n \twoheadrightarrow f$ and the range of f_n is contained in $\overline{\text{co}}\, f(\bar{D})$ for all n. Define

$$d(\phi, D, p) = \lim_{n\to\infty} d(\phi_n, D, p), \tag{7.3.2}$$

where $\phi_n = I - f_n$, and $d(\phi_n, D, p)$ is the Leray–Schauder degree.

Justification (1) Theorem 7.3.3 guarantees the existence of a sequence (f_n) as described.

(2) The Leray–Schauder degree $d(\phi_n, D, p)$ is defined if $p \notin \phi_n(\partial D)$; we show that this is so for large n. Suppose, if possible, that (k_n) is a sequence of integers such that

$$x_{k_n} - f_{k_n}(x_{k_n}) = p \quad (x_{k_n} \in \partial D).$$

Since $\overline{\text{co}}\, f(\bar{D})$ is compact, $(f_{k_n}(x_{k_n}))$ has a convergent subsequence – which, as usual, we take to be the whole sequence: let $f_{k_n}(x_{k_n}) \to \eta$. Then $x_{k_n} \to p + \eta \in \partial D$. It is now easily seen that, because $f_n \twoheadrightarrow f$, $\eta \in f(p+\eta)$, whence $p \in \phi(\partial D)$, a contradiction.

(3) Finally we show that $(d(\phi_n, D, p))$ is an eventually constant sequence; the same argument proves that $d(\phi, D, p)$ as given by (7.3.2) is independent of the particular sequence (ϕ_n) used in its definition. Consider the homotopy

$$H_{n,m}(t, x) = t\phi_n(x) + (1-t)\phi_m(x) \quad (0 \leqslant t \leqslant 1,\ x \in \bar{D}).$$

If $p \in \bigcup_{0\leqslant t\leqslant 1} H_{n,m}(t, \partial D)$ for infinitely many n and m, there are sequences $(x_n) \subset \partial D, (t_n) \subset [0, 1]$ and sequences (k_n), (l_n) of integers such that

$$t_n\phi_{k_n}(x_n) + (1-t_n)\phi_{l_n}(x_n) = p.$$

Now $\overline{\text{co}}\, f(\bar{D})$ is compact, so $(f_{k_n}(x_n))$ and $(f_{l_n}(x_n))$ contain convergent subsequences; we suppose that $f_{k_n}(x_n) \to y_1$ and $f_{l_n}(x_n) \to y_2$. Also we may suppose that $t_n \to t_0$, say. Then $x_n \to x_0 \in \partial D$, say, and $t_0 y_1 + (1-t_0) y_2 = x_0 - p$. But $y_1, y_2 \in f(x_0)$ (because $f_n \twoheadrightarrow f$), so

$$p = t_0(x_0 - y_1) + (1-t_0)(x_0 - y_2) \in t_0\phi(x_0) + (1-t_0)\phi(x_0).$$

Since $\phi(x_0)$ is convex, it follows that $p \in \phi(x_0) \subset \phi(\partial D)$, contrary to hypothesis. By the homotopy invariance of the degree for compact mappings, we deduce that $d(\phi_n, D, p) = d(\phi_m, D, p)$ if n and m are large enough.

We can now prove some of the properties usually associated with degree. The present work does not as it stands fit into the axiomatic scheme presented in Chapter 5. We continue to suppose that D is an open, bounded subset of X and that all mappings occurring are members of $\kappa(\bar{D})$.

Theorem 7.3.5 (1) *If $\phi_n \twoheadrightarrow \phi$ and $p \notin \phi(\partial D)$, then $d(\phi_n, D, p) \to d(\phi, D, p)$.*

(2) *If ψ is a single-valued compact mapping with $\psi(x) \in \phi(x)$ for all $x \in \bar{D}$, then $d(\phi, D, p) = d(\psi, D, p)$ whenever $p \notin \phi(\partial D)$.*

Proof (1) If n is large enough, certainly $p \notin \phi_n(\partial D)$. For $n = 1, 2, \ldots$, take sequences (ϕ_{nm}) of compact single-valued mappings on $\bar{D}$ such that $\phi_{nm} \twoheadrightarrow \phi_n$ as $m \to \infty$ and $d(\phi_n, D, p) = \lim_{m \to \infty} d(\phi_{nm}, D, p)$. Then there is a sequence $(m(n))$ such that $\phi_{nm(n)} \twoheadrightarrow \phi$ as $n \to \infty$ and $d(\phi_{nm(n)}, D, p) = d(\phi_n, D, p)$. Then $d(\phi, D, p) = \lim_{n \to \infty} d(\phi_{nm(n)}, D, p) = \lim_{n \to \infty} d(\phi_n, D, p)$.

(2) This follows from (1) by taking $\phi_n = \psi$ for all n.

Theorem 7.3.6 *If $d(\phi, D, p) \neq 0$, then there is $x \in \bar{D}$ such that $p \in \phi(x)$.*

Proof Suppose that ϕ_n are single-valued compact mappings such that $\phi_n \twoheadrightarrow \phi$ and $d(\phi_n, D, p) \to d(\phi, D, p)$ as $n \to \infty$. For sufficiently large n, $d(\phi_n, D, p) \neq 0$, whence there exists $x_n \in D$ such that $\phi_n(x_n) = p$. The sequence $(x_n - \phi_n(x_n))$ is contained in the compact set $\overline{\text{co}}\, f(\bar{D})$ by construction; it therefore has a convergent subsequence. So $x_n \to \xi$, say, and $p \in \phi(\xi)$.

Theorem 7.3.7 (1) *Let $H = I - h$, where $h: [0, 1] \times \bar{D} \to \Gamma(X)$ is compact. If $p \notin H(t, \partial D)$ for $0 \leqslant t \leqslant 1$, then $d(H(t, .), D, p)$ is independent of $t \in [0, 1]$.*

(2) *Let D be the disjoint union of open sets $D_i\,(i = 1, \ldots, n)$. Then, if $p \notin \phi(\partial D)$,*

$$d(\phi, D, p) = \sum_{i=1}^{n} d(\phi, D_i, p).$$

We leave the proofs of the two parts of Theorem 7.3.7 to the reader, and invite him also to derive in this context the appropriate versions of some of the other properties of degree which we have encountered elsewhere.

To end this section we indicate how Petryshyn and Fitzpatrick (1974) extended the degree defined above to a wider class of multivalued mappings. We still suppose that ϕ maps $\bar{D}$ into $\Gamma(X)$. A transfinite sequence D_α is defined inductively. Let $D_0 = \overline{\text{co}}\,\phi(\bar{D})$. If α is an ordinal of the first kind, define $D_\alpha = \overline{\text{co}}\,\phi(\bar{D} \cap D_{\alpha-1})$; if α is an ordinal of the second kind, let $D_\alpha = \bigcap_{\beta<\alpha} D_\beta$. Then each D_α is closed and convex, $D_\alpha \subset D_\beta$ if $\alpha \geqslant \beta$, and $\phi(D_\alpha \cap \bar{D}) \subset D_\alpha$ for every ordinal α. The sequence D_α is non-increasing, so $D_\gamma = D_{\gamma+1}$ for some ordinal γ. Let $D^* = D_\gamma$; it is easily seen that $\overline{\text{co}}\,\phi(D \cap D^*) = D^*$.

Definition 7.3.8 (1) The mapping $\phi: \bar{D} \to \Gamma(X)$ is *ultimately compact* if it is upper semicontinuous and D^* is compact or $D^* \cap D = \emptyset$.

(2) If ϕ is ultimately compact and $p \notin \phi(\partial D)$, define

$$d(I \dot{-} \phi, D, p) = d(I - \phi \circ r, r^{-1}(D), p), \tag{7.3.3}$$

where r is a retraction of X onto D^*, and the degree on the right hand side is as defined above.

This definition requires some justification. It is necessary to show that the right hand side of (7.3.3) is defined, and is independent of the retraction used. The details of the checking are omitted, as are the derivations of the usual properties of degree. The collection of ultimately compact multivalued mappings is substantially larger than the set of compact mappings, and contains the analogues of strict set contractions, strict ball contractions and condensing maps. The relation between ultimately compact and compact multivalued mappings is strikingly similar to that between k-set contractions and single-valued compact mappings. Webb (1974, 1975) also defines the degree of multivalued mappings such as those considered above, and proves a uniqueness result using axioms similar to those of Chapter 5.

8
Differentiable mappings

8.1 Calculation of degree

We saw in Theorem 2.2.3 that in finite dimensional spaces, the index of an isolated solution x_0 of $\phi(x) = p$ when ϕ is differentiable and p is a regular value of ϕ is simply $(-1)^\nu$, where ν is the sum of the multiplicities of the negative, real eigenvalues of $\phi'(x_0)$. Clearly this is a useful aid to calculating the degree of a mapping. We shall show that the formula holds quite generally, exploiting the ideas of Chapter 5. A proof of this for the Leray–Schauder degree may be found in the book of Krasnosel'skii (1964); it was proved for the degree of mappings of the form 'identity + strict set contraction' by Stuart and Toland (1973), and Thomas (1973). A similar result for the degree of A-proper mappings was given by Fitzpatrick (1972).

Suppose that X is a Banach space and ω the collection of bounded, open subsets of X. Let $M(\omega)$ be an admissible class of mappings and d a topological degree for $M(\omega)$; we suppose that $M(D)$ is convex for each $D \in \omega$. We consider a differentiable mapping $\phi \in M(D)$ and an isolated solution x_0 of $\phi(x) = p$.

We make the following hypotheses: (1) $T = \phi'(x_0) \in M(D)$, (2) T is bijective, (3) the spectrum of T in $(-\infty, 0)$ consists of a finite number of points, each of finite multiplicity. We shall also need a technical hypothesis which, because of its nature, we introduce in the course of the argument.

It was shown in Theorem 5.2.3 that

$$i(\phi, x_0, p) = i(T, x_0, y_0),$$

where $y_0 = T(x_0)$. Without loss of generality, we may take $x_0 = 0$. Let E be the vector space spanned by the set

$$\{x; (T - \lambda I)^k x = 0, \text{ some } k > 0 \text{ and } \lambda < 0\}.$$

By hypothesis, E is finite dimensional; therefore E has a topological complement in $X: X = E \oplus F$, say. Clearly E is invariant under T. We make the additional hypothesis: (4) F can be so chosen that it is invariant under T. Let $\hat{T} = T|_E$, and define

$$h(t) = (1 - t)(\hat{T} \oplus I_F) + tT \quad (0 \leqslant t \leqslant 1),$$

where the notation is as used in Chapter 5. If $h(t)(x) = 0$, with $x = y + z\,(y \in E, z \in F)$, then $T(y) = 0$ and $(1 - t)z + tT(z) = 0$. Using hypotheses (2) and (3), and the definition of E, we deduce that $y = 0$ and $z \in E$; thus $z = 0$ and $x = 0$. It follows that $i(T, 0, 0) = i(\hat{T} \oplus I_F, 0, 0)$. By (5.2.4), this is $i(\hat{T}, 0, 0)$. But by Theorem 5.2.6, $i(\hat{T}, 0, 0) = (-1)^\beta$ where β is the sum of the multiplicities of the negative eigenvalues of $\hat{T}$. However $(-1)^\beta = (-1)^\nu$, where ν is the corresponding sum for T.

Theorem 8.1.1 *Under the above conditions,* $i(\phi, x_0, p) = (-1)^\nu$, *where* ν *is the sum of the multiplicities of the negative eigenvalues of* T.

Remark Nussbaum (1971*a*) showed that all our hypotheses are satisfied for $\phi \in \Sigma_k(\bar{D})$ if $k < 1$.

There is a global version of Theorem 8.1.1; it follows as in Section 5.2.

Corollary 8.1.2 *Suppose that* $D \in \omega$ *contains finitely many solutions of* $\phi(x) = p$. *If the hypotheses of Theorem* 8.1.1 *hold at each point of* $\phi^{-1}(p)$, *then*

$$d(\phi, D, p) = \sum_{x \in \phi^{-1}(p)} (-1)^{\nu(x)}, \tag{8.1.1}$$

where $\nu(x)$ *has the obvious meaning.*

Before proceeding further we introduce some concepts from global analysis. We give the definition in terms of Banach spaces X and Y (rather than Banach manifolds).

Definition 8.1.3 The linear operator $T: X \to Y$ is a *Fredholm operator* if T is closed and both ker T and coker T have finite dimension. The *index* of T, Ind T, is dim ker T – dim coker T.

In this definition ker T is the kernel of T and coker $T = Y/\text{Im } T$, Im T being the range of T. This use of the word 'index' is unfortunate in the context of this book, but unavoidable; no ambiguity is likely.

Definition 8.1.4 Suppose that $\phi:\bar{D} \to Y$ is continuously differentiable. If, for all $x \in D, \phi'(x)$ is a Fredholm operator of index k, then ϕ is said to be *Fredholm of index k*.

We shall write $\Phi_k(D)$ for the collection of Fredholm mappings of $\bar{D}$ into X of index k. A point p is a *regular value* of ϕ if $\phi'(x)$ is surjective for all $x \in \phi^{-1}(p)$. If p is a regular value and $\phi \in \Phi_0(D)$, then dim coker T = dim ker $T = 0$, whence $\phi'(x)$ is bijective for all $x \in \phi^{-1}(p)$. Thus Theorem 8.1.1 is particularly useful when dealing with Fredholm mappings of index zero. A differentiable strict set contraction is known to be Fredholm of index zero.

8.2 Another definition of degree

The formula (8.1.1) can be used as the starting point of a definition of degree for certain mappings defined on subsets of a Banach space. This approach was used by Fenske (1971) to define the degree of elements of $\Sigma_k(\bar{D})(k < 1)$, and was further exploited by Dancer (1975). In this section we very briefly describe the method, which is essentially that used to define the Brouwer degree. We need the following extension of Sard's theorem, proved by Smale (1965); we adapt the version quoted by Elworthy and Tromba (1970) to suit our needs. The reader is referred to Smale's paper for a proof. We retain the notation of Section 8.1.

Theorem 8.2.1 *Suppose that $\phi:\bar{D} \to Y$ is C^r, proper, and Fredholm of index k. If $r >$ max$(k, 0)$, then the set of regular values of ϕ is open and dense in Y.*

Let $\mathscr{F}(D)$ denote the collection of mappings $\phi = I - f:\bar{D} \to X$ which are such that (1) ϕ is C^2 (that is, twice continuously differentiable), (2) ϕ is proper, and (3) $\lambda I - f'(x) \in \Phi_0(D)$ for $\lambda \geqslant 1$ and $x \in D$. If $\tilde{f}'$ is the complexification of f' (defined on the complexification $\tilde{X}$ of X), it is shown by Gokhberg and Krein (1960) that condition (3) in fact implies the following: (3′) for $x \in D$, the spectrum $\sigma(\tilde{f}'(x))$ of $\tilde{f}'(x)$ in $[1, \infty)$ consists of a finite number of points each of which is isolated in $\sigma(\tilde{f}'(x))$ and of finite multiplicity.

It is clear that $\{\mathscr{F}(D); D \in \omega\}$ is an admissible class of mappings (in the nomenclature of Chapter 5); we define the degree of $\phi \in \mathscr{F}(D)$ as follows.

Definition 8.2.2 Suppose that $\phi = I - f \in \mathscr{F}(D)$ and $p \notin \phi(\partial D)$. If p is a regular value of ϕ, define

$$d(\phi, D, p) = \sum_{x \in \phi^{-1}(p)} (-1)^{\nu(x)}, \tag{8.2.1}$$

where $\nu(x)$ is the sum of the multiplicities of the eigenvalues of $f'(x)$ in $(1, \infty)$. If p is not a regular value of ϕ, choose a sequence (p_n) of regular values such that $p_n \to p$; define $d(\phi, D, p)$ to be $\lim_{n \to \infty} d(\phi, D, p_n)$.

This definition requires a considerable amount of justification. For $x \in \phi^{-1}(p)$, condition (3′) ensures that $\nu(x)$ is defined. Since ϕ is proper, $\phi^{-1}(p)$ is compact. Because ϕ is Fredholm of index zero, $\phi'(x)$ is injective when it is surjective, whence $\phi^{-1}(p)$ is a set of isolated points when p is a regular value. Consequently $\phi^{-1}(p)$ is finite, and the summation in (8.2.1) is finite. As for the definition of $d(\phi, D, p)$ when p is not regular, Theorem 8.2.1 certainly implies that a sequence such as (p_n) exists; we shall see later that the sequence $(d(\phi, D, p_n))$ is eventually constant. The following lemma is the starting point. We write $L(X)$ for the collection of continuous linear operators on X. If $T \in L(X)$, we write $\tilde{T}$ for its complexification, $\sigma(T)$ for its spectrum and $\sigma(\tilde{T})$ for the spectrum of $\tilde{T}$. (For an explanation of complexification, see, for example, Chapter 1 of Rickart (1960).) We say that $T \in L_0(X)$ if $\sigma(\tilde{T}) \cap [1, \infty)$ consists of a finite number of points each of which is isolated in $\sigma(\tilde{T})$ and of finite multiplicity; if $\nu(T)$ is the sum of these multiplicities, we write $\gamma(T) = (-1)^{\nu(T)}$.

Lemma 8.2.3 *Suppose that $T \in L_0(X)$ and $1 \notin \sigma(T)$. There is $\delta > 0$ such that if $T_1 \in L(X)$ and $\| T - T_1 \| < \delta$, then $T_1 \in L_0(X)$, $1 \notin \sigma(T_1)$ and $\gamma(T) = \gamma(T_1)$.*

Proof Let $\sigma(T) \cap (1, \infty) = \{\lambda_1, \dots, \lambda_k\}$. Choose $\varepsilon > 0$ so that the discs $B(1, \varepsilon)$, $B(\lambda_1, \varepsilon), \dots, B(\lambda_k, \varepsilon)$ are disjoint and do not meet $\sigma(\tilde{T}) \backslash \sigma(T)$. Let π_i be the circle of radius $\frac{1}{2}\varepsilon$ with centre λ_i; then

$$v(T) = \sum_{i=1}^{k} v(\lambda_i, T), \quad \text{where} \quad v(\lambda_i, T) = \dim E(\lambda_i, \tilde{T})(\tilde{X})$$

and

$$E(\lambda_i, \tilde{T}) = \frac{1}{2\pi i} \int_{\pi_i} (\lambda I - \tilde{T})^{-1} \, d\lambda.$$

(See, for example, Rudin (1973), Chapter 10.) So choose $\delta_1 > 0$ that if $\| T - T_1 \| < \delta_1$, then $\sigma(\tilde{T}_1)$ lies in the $\frac{1}{2}\varepsilon$-neighbourhood of $\sigma(\tilde{T})$. Let $\sigma_i = \sigma(\tilde{T}_1) \cap B(\lambda_i, \frac{1}{2}\varepsilon)$ $(i = 1, \dots, k)$. Now choose $\delta < \delta_1$ so that, if $\| T - T_1 \| < \delta$, then $v(\lambda_i, T) = v_i(T_1) (i = 1, \dots, k)$; here

$$v_i(T_1) = \sum_{\lambda_{ij} \in \sigma_i} v(\lambda_{ij}, T_1) = \sum_{\lambda_{ij} \in \sigma_i} \dim E(\lambda_{ij}, \tilde{T}_1)(\tilde{X}).$$

But $\bar{\mu} \in \sigma(\tilde{T}_1)$ if $\mu \in \sigma(\tilde{T}_1)$, and $v(\mu, T_1) = v(\bar{\mu}, T_1)$. Hence $v_i(T_1) - \dim E(\sigma_i \cap \mathbf{R}, T_1)(\tilde{X})$ is an even number. Therefore $\gamma(T) = \gamma(T_1)$, as required.

Remark Underlying Lemma 8.2.3 is the idea that a real eigenvalue of a real operator 'loses' multiplicity under perturbation by the appearance of complex eigenvalues in conjugate pairs.

Theorem 8.2.4 *Suppose that $H = I - h: \bar{D} \times [0, 1] \to X$ is twice continuously differentiable and proper, and that the partial derivative $\partial_x h(x, t)$ always satisfies condition (3′) of the above. If, for $(x, t) \in \partial D \times [0, 1]$, $H(x, t) \neq p$, and p is a regular value for $\phi(x) = H(x, 0)$ and $\psi(x) = H(x, 1)$, then $d(\phi, D, p) = d(\psi, D, p)$.*

Proof There is certainly a neighbourhood U of p such that (*a*) U is disjoint from $H(\partial D \times [0, 1])$, (*b*) if $z \in U$, z is regular for both ϕ and ψ, and (*c*) $d(\phi, D, z) = d(\phi, D, p)$ and $d(\psi, D, z) = d(\psi, D, p)$ for all $z \in U$. Now H is C^2 and is a Fredholm mapping of index 0 or 1; so, by Theorem 8.2.1, the regular points of H are dense in X. We can therefore suppose, without loss of generality, that p itself is regular for ϕ, ψ and H. In that case $H^{-1}(p)$ is a compact one-dimensional manifold, and so is the disjoint union of paths. Since $p \notin H(\partial D, [0,1])$, these paths start and end in $D \times \{0\}$ or $D \times \{1\}$, without meeting $\partial D \times [0,1]$. Take $x_0 \in \phi^{-1}(p)$; there is a differentiable path in $H^{-1}(p)$ joining x_0 to a point x_1 of $\phi^{-1}(p)$ or of $\psi^{-1}(p)$. We prove that in the former case, $\nu(f'(x_0)) = \nu(f'(x_1)) + 1$ (mod 2), and in the latter $\nu(f'(x_0)) = \nu(g'(x_1))$ (mod 2); here $f = I - \phi$ and $g = I - \psi$.

We consider the case $x_1 \in \phi^{-1}(p)$. Let c be a path in $H^{-1}(p)$ with $c(0) = (x_0, 0)$ and $c(1) = (x_1, 0)$. For $t \in [0,1]$, let $K_t = \ker H'(c(t))$; there is a decomposition $X \times \mathbf{R} = X_t \times K_t$, where $H'(c(t))$ is a bijection of X_t onto X. If $y \in D \times I$, write $y = (\xi, \tau)$, where $\xi \in X_t$ and $\tau \in K_t$. Define $H_t(y) = (H(y), \tau)$; H_t is non-singular at $c(t)$. There is therefore a chain $0 = t_0 < t_1 < \ldots < t_n = 1$ such that $H_i \equiv H_{t_i}$ is a diffeomorphism on a neighbourhood U_i of $c(t_i)$, and $U_i \cap U_{i+1} \cap c \neq \emptyset$. Write $H_i = I - h_i$. If $c(s) \in U_i \cap U_{i+1}$, then clearly

$$\nu(h_i'(c(s))) = \nu(h_{i+1}'(c(s))).$$

By Lemma 8.2.3 and a compactness argument,

$$\nu(h_i'(c(t_i))) = \nu(h_i'(c(s))) \pmod 2$$

and
$$\nu(h_{i+1}'(c(t_{i+1}))) = \nu(h_{i+1}'(c(s))) \pmod 2.$$

Hence
$$\nu(h_i'(c(t_i))) = \nu(h_{i+1}'(c(t_{i+1}))) \pmod 2.$$

It follows that
$$\nu(h_0'(c(0))) = \nu(h_n'(c(1))) \pmod 2.$$

For $i = 0, n$, we decompose $X \times \mathbf{R}$ as $X \times T_i$, where T_i is tangential to c at $c(t_i)$; let k_i be the restriction of h_i to T_i. Then, if $z_0 = x_0$ and $z_n = x_1$,

$$h_i'(c(t_i)) = (f'(z_i), k_i'(c(t_i))) \quad (i = 0, n),$$

whence
$$\nu(h_i'(c(t_i))) = \nu(f'(z_i)) + \nu(k_i'(c(t_i))).$$

Since p is regular for $H, I - k_i'(c(t_i))$ $(i = 0, n)$ are both non-singular mappings of $\mathbf{R}$ into $\mathbf{R}$, and their eigenvalues have opposite signs.

Thus one and only one of the mappings $k_0'(c(0))$ and $k_n'(c(1))$ has an eigenvalue in $(1, \infty)$. So

$$v(k_0'(c(0))) = -v(k_n'(c(1))) \pmod 2.$$

Hence $$v(f'(x_0)) = -v(f'(x_1)) \pmod 2.$$

The case of $x_1 \in \psi^{-1}(p)$ can be treated similarly, so completing the proof.

We now use this restricted version of homotopy invariance to deduce a theorem which in turn finally justifies the second part of our definition of degree.

Theorem 8.2.5 *Suppose that p_1 and p_2 are regular values of $\phi = I - f \in \mathscr{F}(D)$ which lie in the same component of $\mathscr{C}\phi(\partial D)$. Then $d(\phi, D, p_1) = d(\phi, D, p_2)$.*

Proof Let c be a path in X not meeting $\phi(\partial D)$ and such that $c(0) = p_1$, $c(1) = p_2$; let $\eta = \frac{1}{2}\rho(c, \phi(\partial D))$. If $c(t)$ is not a regular value of ϕ, choose ξ_t with $\|\xi_t\| < \eta$ so that $c(t) - \xi_t$ is regular (this is an application of Theorem 8.2.1); if $c(t)$ is already a regular value, take $\xi_t = 0$. Define $F_t(x) = \phi(x) + \xi_t$; for all $t \in [0, 1]$, $c(t)$ is a regular value of F_t. There is $\eta_t (0 < \eta_t < \eta)$ such that $B(t) = B(c(t), \eta_t)$ contains only regular values of F_t. A chain $0 = t_0 < t_1 < \ldots < t_n = 1$ can be found so that $c \subset \bigcup B(t_i)$ and $c \cap B(t_i) \cap B(t_{i+1}) \neq \emptyset$. Suppose that $c(\tau) \in B(t_i) \cap B(t_{i+1})$. Then $d(F_{t_i}, D, c(t_i)) = d(F_{t_i}, D, c(\tau))$ and $d(F_{t_{i+1}}, D, c(t_{i+1})) = d(F_{t_{i+1}}, D, c(\tau))$. Now consider the homotopy $H: \bar{D} \times I \to X$ given by

$$H(x, \lambda) = x - f(x) + (1 - \lambda)\xi_{t_{i+1}} + \lambda\xi_{t_i} \quad (x \in \bar{D}, 0 \leqslant \lambda \leqslant 1).$$

If $H(x, \lambda) = c(\tau)$ with $x \in \partial D$ and $0 \leqslant \lambda \leqslant 1$, then

$$\|x - f(x) - c(\tau)\| \leqslant (1 - \lambda)\|\xi_{t_{i+1}}\| + \lambda\|\xi_{t_i}\| < \eta,$$

contradicting the choice of η. By Theorem 8.2.4, $d(F_{t_i}, D, c(\tau)) = d(F_{t_{i+1}}, D, c(\tau))$. It follows that $d(F_0, D, c(0)) = d(F_1, D, c(1))$. However, $p_1 = c(0)$ and $p_2 = c(1)$ are both regular values of ϕ, and $F_0 = F_1 = \phi$. The proof is complete.

It is now clear that Definition 8.2.2 is fully justified. It is also not difficult to see that d satisfies the axioms for a topological degree (the reader is left to supply the remaining details that are needed). Hence d has all the properties of a degree which were deduced in Chapter 5. The mappings considered in Chapter 6 are known to have the defining properties of $\mathscr{F}(D)$; Theorem 6.2.6 tells us, therefore, that the two degrees we have now defined for such mappings coincide. We note that Elworthy and Tromba (1970) have used methods similar to those of this section to define a degree for mappings between Banach manifolds; this, however, is beyond the scope of this book. Thomas (1973) proved the result of Theorem 8.1.1 for differentiable mappings of the form 'Identity + strict set contraction'; he deduced the uniqueness of the degree of such mappings using an axiom scheme somewhat different from that of Chapter 5.

Note Recently Rothe has developed an approach to the definition of Leray–Schauder degree very much in the spirit of this chapter. His method, functional analytic in nature, defines Leray–Schauder degree without using the Brouwer degree. The details are as yet unpublished; they are dependent on an earlier paper (E. Rothe: Mapping degree in Banach spaces and spectral theory. *Math. Z.* **63**, (1955), 195–218).

9
Some applications of degree theory

In this, the final chapter, some of the applications of degree theory in analysis are surveyed. The selection of topics inevitably reflects, at least partly, the predilections of the author. The choice cannot in any sense be complete, nor can any particular field be discussed in detail without seriously distorting the balance of the book. It may be hoped, however, that the reader will acquire some feeling for the techniques used; to assist in this, the accent will be on the particular rather than the general. Some applications have been encountered in earlier chapters – fixed point theorems, especially. The emphasis in the chapter as a whole will be on the rôle of degree theory in the study of ordinary differential equations; the examples have been chosen for illustrative purposes.

9.1 Periodic solutions (I)

We start by establishing some notation and terminology. The equation

$$\frac{dx}{dt} \equiv \dot{x} = f(x,t) \quad (x \in \mathbf{R}^n, t \in \mathbf{R}) \tag{9.1.1}$$

is said to be *admissible* if $f: \mathbf{R}^n \times \mathbf{R} \to \mathbf{R}^n$ is continuous and, given t_0 and x_0, there is one and only one solution $\phi(t)$ of (9.1.1) satisfying $\phi(t_0) = x_0$. This solution will be written $x_f(t;t_0,x_0)$ – or simply $x(t;t_0,x_0)$ if no confusion thereby arises. The equation (9.1.1) is certainly admissible if f satisfies a Lipschitz condition in x, with constant independent of t, in every compact subset of $\mathbf{R}^n \times \mathbf{R}$. If (9.1.1) is admissible, then by a standard theorem on differential equations (see, for example, Coddington and Levinson (1955), Chapter 1) $x_f(t;t_0,x_0)$ is, for fixed t, a continuous function of t_0 and x_0. We shall extend the terminology somewhat, and say that

f is admissible whenever (9.1.1) is an admissible equation. Let

$$\mathscr{A} = \{f:\mathbf{R}^n \times \mathbf{R} \to \mathbf{R}^n; f \text{ is admissible}\};$$

the set $\mathscr{A}$ is given the topology of uniform convergence on compact sets. Then $x_f(t;t_0,x_0)$ is a continuous function of f, as well as of t_0 and x_0. On occasion the function f depends on a parameter (ε, say) as well as on x and t. If f is a continuous function of ε, then $x_f(t,t_0,x_0)$ also depends continuously on ε; we shall in these circumstances denote the solutions by $x_f(t;t_0,x_0,\varepsilon)$.

The equation (9.1.1) is said to be *autonomous* if f does not depend explicitly on t; otherwise it is *non-autonomous*. The equation is *periodic* if, for some $\omega > 0$, $f(x,t+\omega) = f(x,t)$ for all x and t. In this section we shall be interested in the existence of periodic solutions of (9.1.1); this is a sensible question only when the equation is either autonomous or periodic. If f has period ω in t, it is easy to show that, if $\phi(t) = x_f(t;0,c)$, then $\phi(t+\omega) = \phi(t)$ for all t if and only if $\phi(\omega) = \phi(0)$. Periodic solutions of period $k\omega (k \in \mathbf{Z})$ are therefore determined by the fixed points of the 'translation map'

$$c \mapsto x_f(k\omega;0,c).$$

We shall look at the zero sets of the mappings $q(\tau,f;.):\mathbf{R}^n \to \mathbf{R}^n$ defined by

$$q(\tau,f;c) = x_f(\tau;0,c) - c \quad (\tau \neq 0, f \in \mathscr{A}). \tag{9.1.2}$$

An important point to consider always is the domain of definition of the function q. The solution $x_f(t;t_0,x_0)$ is only guaranteed to to be defined on some t-interval containing t_0; if its maximal interval of definition is (α,β), then $|x_f(t;t_0,x_0)| \to \infty$ as $t\uparrow\beta$ if β is finite, and as $t\downarrow\alpha$ if α is finite.

In all these remarks it has been supposed that the function f is defined on all of $\mathbf{R}^n \times \mathbf{R}$. It is not difficult to see how everything should be adapted to apply to the case when f is given on a subset $D \times I$ of $\mathbf{R}^n \times \mathbf{R}$, where $D \subset \mathbf{R}^n$ and $I \subset \mathbf{R}$ are both open. The only substantive change is that a solution may approach the boundary of $D \times I$ at a finite end of its maximal interval of definition, rather than tend to infinity.

We can now start looking at particular classes of equations. The equations for which the most complete results can be obtained are those termed quasilinear. Let us consider the equation

$$\dot{x} = f(x, t, \varepsilon) = A(t)x + p(t) + \varepsilon g(x, t, \varepsilon) \quad (x \in \mathbf{R}^n), \qquad (9.1.3)$$

where the components of the matrix A and of the functions p and g are continuous and of period ω in t, ε is a real parameter and g is continuously differentiable in x and ε; ω is independent of ε. f is certainly admissible. We suppose that the equation is strictly non-autonomous; we can then sensibly pose the question: does (9.1.3) have solutions of period ω for sufficiently small ε? (For autonomous equations the possible periods of periodic solutions are unknown *a priori*.)

Let $M(t)$ be a fundamental matrix of the equation

$$\dot{y} = A(t)y \qquad (9.1.4)$$

– that is, $M(t)$ is non-singular (for all t), $dM/dt = AM$ and $M(0) = I$, the identity matrix. By the formula of variation of constants, the solutions of (9.1.3) are given by

$$x_f(t;0,c,\varepsilon) = M(t)c + M(t)\int_0^t [M(s)]^{-1}[p(s) + \varepsilon g(x(s;0,c,\varepsilon),s,\varepsilon)]\,ds.$$

The zeros of $q(\omega, f; c)$ are given by

$$(M(\omega) - I)c + M(\omega)\int_0^\omega [M(s)]^{-1}[p(s) + \varepsilon g(x(s;0,c,\varepsilon), s, \varepsilon)]ds = 0. \qquad (9.1.5)$$

Our task is to solve this equation for c in terms of ε.

When the matrix $M(\omega) - I$ is non-singular, we have what is usually called the non-resonance case; it occurs when (9.1.4) has no non-trivial solution of period ω – so that the period of the 'forcing term' g is not a natural period of that equation. In this case, the implicit function theorem yields the existence of a unique solution of (9.1.5) when ε is sufficiently small:

$$c = -[M(\omega) - I]^{-1}M(\omega)\int_0^\omega [M(s)]^{-1}p(s)\,ds + O(\varepsilon).$$

When $M(\omega) - I$ is singular, the position is a good deal more complicated. We shall consider only the completely degenerate case – when $M(\omega) = I$. Equation (9.1.5) can be solved for all small ε only if

$$\int_0^{\omega} [M(s)]^{-1} p(s)\, ds = 0. \tag{9.1.6}$$

It is then required to solve the equation

$$N_\varepsilon(c) = \int_0^{\omega} [M(s)]^{-1} g(x(s;0,c,\varepsilon), s, \varepsilon)\, ds = 0$$

for small ε. The following is a straightforward application of Theorems 2.1.1 and 2.1.2.

Lemma 9.1.1 *Suppose that D is a bounded, open subset of* $\mathbf{R}^n$ *such that $d(N_0, D, 0) \neq 0$. Then, if* (9.1.6) *is satisfied,* (9.1.3) *has a solution of period ω for all small enough ε.*

As it stands, Lemma 9.1.1 is not very useful unless $d(N_0, D, 0)$ can be calculated. In certain cases it is possible to calculate this degree explicitly (using the techniques of Chapter 8, for example); more usually it is shown that N_0 is homotopic either to the identity or to an odd mapping.

For a detailed account of the use of degree in 'small parameter theory' the reader is referred to Cronin (1964). We illustrate the use of Lemma 9.1.1 by considering the problem of 'entrainment of frequency' – a question of interest in the physical applications of differential equations. Consider the two-dimensional system

$$\left.\begin{aligned} \dot{x} &= y + \varepsilon(h_1(x, y, \varepsilon) + k_1(\tau)) \\ \dot{y} &= -x + \varepsilon(h_2(x, y, \varepsilon) + k_2(\tau)), \end{aligned}\right\} \tag{9.1.7}$$

where h_1 and h_2 are polynomials in their arguments, k_1 and k_2 are trigonometric polynomials, and

$$\tau = t(1 + \varepsilon\eta(\varepsilon))^{-1};$$

here η is a continuous function with $\eta(0) = \eta_0 \neq 0$. When $\varepsilon = 0$ all solutions of (9.1.7) have period 2π; we ask whether, when $\varepsilon \neq 0$, there are solutions of period $\omega(\varepsilon) = 2\pi(1 + \varepsilon\eta(\varepsilon))$. The forcing term is small and has period close to 2π; the existence of solutions of of period $\omega(\varepsilon)$ means that the forcing term has created oscillations of its period, rather than of the natural period of the unforced system. We use τ as a new independent variable and apply the previous theory to (9.1.7), which now has period 2π in τ. Condition (9.1.6) is certainly satisfied. The equation $N_\varepsilon(c) = 0$ becomes

$$\int_0^{2\pi} \begin{pmatrix} \cos s, & -\sin s \\ \sin s, & \cos s \end{pmatrix} \begin{pmatrix} \eta(\varepsilon)y + (1 + \varepsilon\eta(\varepsilon))(h_1 + k_1) \\ -\eta(\varepsilon)x + (1 + \varepsilon\eta(\varepsilon))(h_2 + k_2) \end{pmatrix} ds = 0.$$

$N_0(c) = 0$ is equivalent to the pair of equations

$$2\pi\eta_0 c_2 + H_1(c_1, c_2) + K_1 = 0$$

$$-2\pi\eta_0 c_1 + H_2(c_1, c_2) + K_2 = 0,$$

where $c = (c_1, c_2)$ is to be found, K_1 and K_2 are constants, and H_1 and H_2 are polynomials in which each term is of odd power. (The details of this calculation may be checked quite easily.) Moreover, N_0 is given by polynomials of odd degree which, because $\eta_0 \neq 0$, do not vanish identically. So $d(N_0, D, 0)$ is defined if D is a sufficiently large disc, and Theorem 3.2.6 implies that $d(N_0, D, 0)$ is odd, and hence non-zero.

Theorem 9.1.2 *Under the stated conditions, there is* $\varepsilon_0 > 0$ *such that* (9.1.7) *has at least one solution of period* $2\pi(1 + \varepsilon\eta(\varepsilon))$ *if* $\varepsilon < \varepsilon_0$.

The successful uses of degree in small parameter theory work because of intermediate results of the type: 'if $\varepsilon < \varepsilon_0$, a certain degree is defined and is non-zero'. There is some hope that on occasion ε_0 need not be small; then the homotopy N_ε of the above joins N_0 to N_1, say, so breaking out of the 'small parameter' restriction. A similar idea to this is to embed a given equation in a family depending on a parameter (μ, say); $\mu = 1$ usually corresponds to the given equation, while $\mu = 0$ corresponds to an equation whose properties are well understood. For this approach to succeed it is usually necessary to impose conditions that enable an *a priori* bound for the size of periodic solutions to be obtained; in this way the degree of the translation map is shown to be defined relative to a suitably large domain which is independent of μ. An interesting example of this technique is the following, given by Cronin (1974); the reader will find references to work along similar lines in that paper.

We consider the equation

$$\dot{x} = f(x,t,\mu) = A(t)x + \mu g(x,t,\mu) \quad (x \in \mathbf{R}^n). \tag{9.1.8}$$

Here μ is a real parameter lying in the interval $[0, 1]$, and the components of the matrix A and the vector g are all of period ω in t and continuously differentiable for all t, x and $\mu \in [0, 1]$. We make the hypotheses:

(1) the equation $\dot{y} = A(t)y$ has no non-trivial solutions of period ω,

(2) there exist positive constants R, N and ν such that $0 \leqslant \nu \leqslant 1$,

$$|g(x,t,\mu)| < N|x|^{1-\nu} \quad (|x| \geqslant R, 0 \leqslant t \leqslant \omega, 0 \leqslant \mu \leqslant 1)$$

and
$$|g(x,t,\mu)| < NR^{1-\nu} \quad (|x| \leqslant R, 0 \leqslant t \leqslant \omega, 0 \leqslant \mu \leqslant 1).$$

As before, we denote the solutions of (9.1.8) by $x(t; t_0, x_0, \mu)$; $x(t; t_0, x_0, \mu)$ is continuous in each argument. Because $f = O(|x|)$, every solution is defined for all t. We again have to solve (9.1.5) (with $p = 0$); by hypothesis (1) above, $M(\omega) - I$ is a non-singular matrix. Equation (9.1.5) is thus of the form $Lc + \mu T(c, \mu) = 0$ where L is linear and non-singular, while T is non-linear. The next step is to bound $|T(c, \mu)|$. The C_i in the following are constants; we write

$$\alpha(c) = \max_{\substack{0 \leqslant t \leqslant \omega \\ 0 \leqslant \mu \leqslant 1}} |x(t; 0, c, \mu)|, \quad \beta(c) = \max_{\substack{0 \leqslant t \leqslant \omega \\ 0 \leqslant \mu \leqslant 1}} |g(x(t; 0, c, \mu), t, \mu)|.$$

It is clear that $|T(c, \mu)| \leqslant C_1 \beta(c)$. If $\alpha(c) < R$, then $\beta(c) < NR^{1-\nu}$; if $\alpha(c) \geqslant R$, $\beta(c) < N(\alpha(c))^{1-\nu}$. It is not difficult to show that $\alpha(c) \leqslant \alpha(0) + C_2|c|$ (using an argument similar to Gronwall's inequality (1919)). We deduce that

$$|T(c, \mu)| \leqslant C_3 N(C_4 + |c|)^{1-\nu}.$$

Theorem 9.1.3 *Under the stated conditions,* (9.1.8) *has a periodic solution for all* $\mu \in (0, 1]$.

Proof We have to solve

$$h_\mu(c) = c + \mu L^{-1} T(c, \mu) = 0.$$

Suppose that $h_\mu(c) = 0$ and $c \neq 0$; then

$$|c| \leqslant \mu \| L^{-1} \| C_3 N (C_4 + |c|)^{1-\nu}.$$

Either $|c| \leqslant C_4$ or $|c|^\nu < C_5 \mu$. We conclude that there is $C_6 > 0$, independent of μ, such that every periodic solution of (9.1.8) is in the ball $B = B(0, C_6)$ when $t = 0$. It follows that the Brouwer degree $d(h_\mu, B, 0)$ is defined for all $\mu \in [0, 1]$, and, by homotopy invariance, equals $d(h_0, B, 0)$. But $h_0 = I$, whence $d(h_\mu, B, 0) = 1$ for all μ. There is therefore a zero of h_μ in B, and this corresponds to a periodic solution of our equation.

Remark So far we have used only the Brouwer degree. Other degrees will be required later.

9.2 Periodic solutions (II)

We now look at differential equations with so-called 'large' nonlinearities. The methods used are not the same as those used in the case of quasilinear equations. Many authors have considered problems in this area – we give a few references: Cronin (1965, 1967), Gomory (1956), Halanay (1966), Lando (1971), Reissig (1970, 1972*a*, 1972*b*).

Definition 9.2.1 A point p is a point of *ω-non-recurrence from* $t = t_0$ for the equation $\dot{x} = f(x, t)$ if $x_f(t; t_0, p) \neq p$ for $t_0 < t \leqslant t_0 + \omega$. (We are not supposing that $x_f(t; t_0, p)$ is continuable to $t = t_0 + \omega$.)

We consider an equation

$$\dot{x} = f(x, t), \tag{9.2.1}$$

where f is a C^1 function defined on $\mathbf{R}^n \times \mathbf{R}$. As yet, f is not supposed periodic. Define vector fields $F: c \mapsto f(c, 0)$ and $V_t: c \mapsto x_f(t; 0, c) - c$; for a given t, V_t may not be defined for all c.

Theorem 9.2.2 *Suppose that D is an open, bounded subset of* $\mathbf{R}^n$ *such that* (i) $f(x,0) \neq 0 (x \in \partial D)$, (ii) $x_f(t;0,c)$ *is defined for* $c \in \bar{D}$ *and* $0 \leqslant t \leqslant \omega$, *and* (iii) *every point of ∂D is a point of ω-non-recurrence from $t = 0$ for* (9.2.1). *Then*

$$d(V_\omega, D, 0) = d(F, D, 0).$$

Proof By (ii) and (iii), $d(V_t, D, 0)$ is defined for all $t \in (0, \omega]$; by (i), $d(F, D, 0)$ is also defined. Since $0 \notin V_t(\partial D)$ for $0 < t \leqslant \omega$, $d(V_t, D, 0)$ is independent of $t \in (0, \omega]$ (homotopy invariance). To connect the degrees of F and V_t we observe that, if $\phi(t)$ is a component of $x_f(t;0,c)$, then, for some $\tau \in (0, t)$,

$$\phi(t) = \phi(0) + t\dot{\phi}(0) + \tfrac{1}{2}t^2\ddot{\phi}(\tau).$$

For $t \in [0, \omega]$ and $c \in \bar{D}$, $x_f(t;0,c)$ is contained in a compact set independent of c; but ϕ is a C^2 function, so $\phi(t) = \phi(0) + t\dot{\phi}(0) + O(t^2)$ as $t \to 0$. Thus

$$t^{-1} V_t(c) = F(c) + o(1) \quad (t \to 0)$$

uniformly for $c \in \bar{D}$. Thus, for small enough t, $F(c)$ and $V_t(c)$ are certainly not in opposition on ∂D; so, by the Poincaré–Bohl theorem (Theorem 2.1.5), $d(F, D, 0) = d(V_t, D, 0)$ for small t. But we have seen that $d(V_t, D, 0) = d(V_\omega, D, 0)$, whence the result.

Corollary 9.2.3 *Suppose that f is of period ω in t. If the hypotheses of Theorem* 9.2.2 *are satisfied and* $d(F, D, 0) \neq 0$, *then* (9.2.1) *has a periodic solution.*

Proof Since $d(F, D, 0) = d(V_\omega, D, 0)$, there is $\gamma \in D$ such that $x_f(\omega; 0, \gamma) = \gamma$. Thus $x_f(t; 0, \gamma)$ is a solution of period ω.

Remarks (1) The point of Theorem 9.2.2 and Corollary 9.2.3 is that the degree of only the known function F need be considered.

(2) If $x_f(t;0,c)$ is not defined for $0 \leqslant t \leqslant \omega$ for all $c \in \bar{D}$, the theorem is not applicable. In such a case, it is useful to seek a C^1 function $H: \mathbf{R}^n \to \mathbf{R}^+$ such that (i) $H(x) = 1$ if $x \in B_r \equiv B(0, r)$ for some $r > 0$, and (ii) $Hf = O(1)$ as $x \to \infty$, uniformly in t. We then look at the equation

$$\dot{x} = f(x,t)H(x). \tag{9.2.2}$$

The effect of H is to slow solutions down outside B_r to such an extent that they cannot have finite escape time. With $g(x,t) = f(x,t)H(x)$, let $W_t(c) = x_g(t;0,c) - c$. Clearly $W_t(c)$ is defined for all t and c. Suppose now that we can find an open subset D of $\mathbf{R}^n$ such that (i) $\bar{D} \subset B_r$, (ii) $f(c,0) \neq 0$ $(c \in \partial D)$, and (iii) every point of ∂D is a point of ω-non-recurrence from $t = 0$ for (9.2.2). Then

$$d(W_\omega, D, 0) = d(F, D, 0).$$

If f is of period ω and $d(F,D,0) \neq 0$, we deduce that (9.2.2) has a solution of period ω: $\xi(t)$, say. Under certain conditions $\xi(t)$ is also a solution of period ω of (9.2.1). For example, suppose that ∂D consists entirely of points of ω-non-recurrence for (9.2.2) from all $t \in [0, \omega]$. Then $\xi(t)$ cannot meet ∂D, whence ξ is contained entirely in D. But $D \subset B_r$, so that $H = 1$ on the path $\xi(t)$; therefore $\xi(t) = x_f(t;0,\xi(0))$.

If, in Theorem 9.2.2, $d(F,D,0) = d \neq 0$, it is not generally possible to make any statements about the number of periodic solutions. In fact, there may only be one, or there may be infinitely many. We shall see later that when considering holomorphic equations (with $x \in \mathbf{C}^n$), we can conclude that if $d(F,D,0) = d$, then there are exactly d periodic solutions, provided that a proper idea of the multiplicity of periodic solutions is introduced. There is no precise analogue for equations in $\mathbf{R}^n$. However, results along the lines of the following theorem are possible; it states that a certain small perturbation of f gives an equation with at least $|d|$ periodic solutions.

Theorem 9.2.4 *Suppose that the hypotheses of Theorem 9.2.2 hold, and that $d(F,D,0) = d \neq 0$. Given $\varepsilon > 0$, there is a function $h(x,t)$ of period ω in t such that $|f(x,t) - h(x,t)| < \varepsilon$ $(0 \leqslant t \leqslant \omega, x \in \bar{D})$, and the number of solutions of period ω of*

$$\dot{x} = h(x,t) \tag{9.2.3}$$

with initial points in D is finite and at least $|d|$.

Proof First, we know that $d(V_\omega, D, 0) = d(F, D, 0) = d$. Choose

$\varepsilon_1 < \varepsilon$ so that $d(V_\omega, D, 0) = d(V_\omega, D, a)$ when $|a| < \varepsilon_1$ (such a choice is possible, by Theorem 2.1.3). Take $a \in B(0, \varepsilon_1)$ to be a non-critical value of V_ω (this is possible by Sard's theorem). Then $D \cap V_\omega^{-1}(a)$ is finite and contains at least $|d|$ points. Now, for $0 \leqslant t \leqslant \omega$, define

$$h(x,t) = f(x + a\sin^2(t\pi/2\omega), t) - \frac{a\pi}{2\omega}\sin(t\pi/\omega).$$

For values of t outside the range $[0, \omega)$, h is defined so as to be of period ω. The function h thus has jump discontinuities; however, it is still possible to discuss (9.2.3) – the solutions will have discontinuities in their derivatives at $t = m\omega$ $(m \in \mathbf{Z})$. Let

$$\hat{V}_t(c) = x_h(t; 0, c) - c.$$

The solutions $x_h(t; 0, c)$ are given by

$$x_h(t; 0, c) = x_f(t; 0, c) - a\sin^2(t\pi/2\omega).$$

Thus $\hat{V}_t(c)$ is defined for all $t \in [0, \omega]$ if $c \in \bar{D}$, and $\hat{V}_\omega(c) = V_\omega(c) - a$. Recalling the choice of a, we have, using Theorem 2.1.6,

$$d = d(V_\omega, D, 0) = d(V_\omega, D, a) = d(\hat{V}_\omega, D, 0).$$

The zeros of $\hat{V}_\omega$ are the a-points of V_ω, whence 0 is a non-critical value of $\hat{V}_\omega$. Thus the number of zeros of $\hat{V}_\omega$ in D is finite and at least $|d|$. Finally, ε_1 can be chosen sufficiently small that $|f(x,t) - h(x,t)| < \varepsilon$ if $x \in \bar{D}$ and $0 \leqslant t \leqslant \omega$. This completes the proof.

If $\phi(t)$ is an isolated ω-periodic solution of (9.2.1) and $\phi(0) = \eta$, the index of η as a zero of V_ω is defined: $i(V_\omega, \eta, 0) = d(V_\omega, B, 0)$, where B is any neighbourhood of η not containing the initial point of another ω-periodic solution. If 0 is not a critical value of V_ω, then $i(V_\omega, \eta, 0) = \pm 1$. We make the following definitions.

Definition 9.2.5 The *index* of the periodic solution $\phi(t)$ is $i(V_\omega, \phi(0), 0)$. The solution $\phi(t)$ is *simple* if $\phi(0)$ is not a critical point of V_ω.

The following result is immediate.

Theorem 9.2.6 *Suppose that the degree $d(V_\omega, D, 0)$ is defined and* *is d. If all its periodic solutions are simple, then the equation* (9.2.1) *has at least* $|d|$ *solutions of period* ω.

The description of a periodic solution ϕ as 'simple' can be interpreted in terms of the variational equation of (9.2.1). We recall that the *variational equation* of (9.2.1) relative to $\phi(t)$ is the linear equation

$$\dot{\xi} = A(t)\xi \quad (\xi \in \mathbf{R}^n), \tag{9.2.4}$$

where $A(t)$ is the matrix $(\partial f_i(\phi(t), t)/\partial x_j)$ the components of which are, of course, continuous functions of period ω.

Theorem 9.2.7 *Suppose that f is a* C^2 *function. A solution* $\phi(t)$ *of* (9.2.1) *of period* ω *is simple if and only if the variational equation of* (9.2.1) *relative to* ϕ *has no non-trivial solutions of period* ω.

Proof Let $\phi(0) = \eta$; in a neighbourhood of η, we can write, for $0 \leqslant t \leqslant \omega$,

$$V_t(c) = -c + \phi(t) + \beta(t)(c - \eta) + O(|c - \eta|^2),$$

where β is a matrix satisfying (9.2.4) and $\beta(0) = I$. Thus

$$V_\omega(c) = (\beta(\omega) - I)(c - \eta) + O(|c - \eta|^2),$$

so that ϕ is simple if and only if the matrix $\beta(\omega) - I$ is non-singular. But $\beta(\omega) - I$ is non-singular if and only if (9.2.4) has no non-trivial solution of period ω (that is, 1 is not a characteristic multiplier of (9.2.4)).

One of the difficulties with Theorem 9.2.2 is that it may be difficult to find appropriate domains D with ∂D consisting entirely of points of ω-non-recurrence. An interesting and useful case when this problem can be solved without difficulty is the following (see Halanay (1966)). Recall that x_0 is a *critical point* of $\dot{x} = f(x)$ if $f(x_0) = 0$. (For a discussion of the pattern of solutions near critical points, see, for example, Nemytskii and Stepanov (1960).)

Theorem 9.2.8 *Consider the equation*

$$\dot{x} = f(x) = f_0(x) + \mu f_1(x, t) \quad (\mu \in \mathbf{R},\ x \in \mathbf{R}^n), \tag{9.2.5}$$

where f_0 and f_1 are C^1 functions of their arguments and f_1 is of period ω in t. Suppose that $x = 0$ is an isolated critical point of the equation $\dot{x} = f_0(x)$ which is not a limit point of closed orbits of periods less than or equal to ω. If the index $i(f_0, 0, 0) \neq 0$, then there is $\mu_0 > 0$ such that (9.2.5) has a solution $\phi_\mu(t)$ of period ω if $0 < \mu < \mu_0$; moreover, $\phi_\mu(t) \to 0$ as $\mu \to 0$.

Proof Let μ and r be sufficiently small that (i) $x_f(t; 0, c)$ is defined for $0 \leqslant t \leqslant \omega$ if $c \in \bar{B}_r \equiv \overline{B(0, r)}$, and (ii) every point of ∂B_r is a point of ω-non-recurrence from $t = 0$ for the equation $\dot{x} = f_0(x)$. Let $W_t(\mu)$ denote the map $c \mapsto x_f(t; 0, c, \mu) - c$. By Theorem 9.2.2, $d(W_\omega(0), B_r, 0) = i(f_0, 0, 0)$ if r is small enough. But, for small enough μ, $d(W_\omega(\mu), B_r, 0) = d(W_\omega(0), B_r, 0)$, simply because solutions depend continuously on the parameter μ. Since $i(f_0, 0, 0) \neq 0$, (9.2.5) has a solution of period ω with initial point in B_r.

Corollary 9.2.9 *The result of the theorem holds for an equation*

$$\dot{x} = f_0(x) + \mu f_1(x, t) + f_2(x, t), \tag{9.2.6}$$

where f_0 and f_1 are as in the theorem, f_2 is C^1 and of period ω in t, and $f_2 = o(f_0)$ uniformly in t as $x \to 0$.

Proof The corollary is proved in much the same way as the theorem. The vector fields $x \mapsto f_0(x)$ and $x \mapsto f_0(x) + f_2(x, 0)$ both have an isolated zero at $x = 0$, and their indices at $x = 0$ are equal (by, for example, the Poincaré–Bohl theorem).

Remarks (1) The corollary applies in particular to an equation of the form (9.2.5) when f_0 is analytic. If we write $f_0 = \sum_{m=k}^{\infty} f_0^{(m)}$, where the components of $f_0^{(m)}$ are homogeneous polynomials of degree m, then it is necessary only to calculate the index $i(f_0^{(k)}, 0, 0)$, which, trivially, equals $i(f_0, 0, 0)$.

(2) Theorem 9.2.8 (or Corollary 9.2.9) may, of course, be applied to each of the critical points of $\dot{x} = f_0(x)$.

Theorem 9.2.10 *Let $x \in \mathbf{R}^2$, and suppose that f_0, f_1 satisfy the hypotheses of Theorem 9.2.8. Suppose also that the equation $\dot{x} = f_0(x)$ has critical points $\xi_1, \ldots, \xi_r$ which are not of centre or centre-focus* *type. If s of $\xi_1, \ldots, \xi_r$ have non-zero index, then* (9.2.5) *has at least s solutions of period ω if μ is small enough.*

Note Critical points of second order systems of autonomous differential equations can be classified in terms of the behaviour of solutions in the neighbourhood of the points (the 'sector theorem' of Lefschetz (1952)).

Proof The result is obtained by applying Theorem 9.2.8 to each of $\xi_1, \ldots, \xi_r$. Of the elementary critical points (node, col, focus, centre), only those of centre type are excluded by the requirements of Theorem 9.2.8; non-elementary critical points of index zero are excluded, as are critical points of centre-focus type (their indices are $+1$).

Remark The proof of Theorem 9.2.10 remains valid for a critical point ξ of centre type for which the periods of orbits sufficiently near ξ exceed ω.

To conclude this section we show how it is sometimes useful to use the Leray–Schauder degree in order to prove the existence of periodic solutions, rather than the Brouwer degree. Hitherto periodic solutions of (9.2.1) have been sought by considering the map $c \mapsto x_f(\omega; 0, c)$, which is defined on a subset of a finite dimensional space. Alternatively, the problem can, under certain circumstances, be posed as an equation in an infinite dimensional function space. Let $\mathscr{B}$ be the Banach space of continuous functions $\mathbf{R} \to \mathbf{R}$ of period ω, with the supremum norm. Given $p \in \mathscr{B}$, denote by $\mathscr{T}(p)$ the collection of periodic solutions of the equation $\dot{x} = f(p(t), t)$. $\mathscr{T}(p)$ will, in general, be a set whose size depends on p, and may be empty. The periodic solutions of (9.2.1) are the fixed points of the multivalued mapping $\mathscr{T}: p \to \mathscr{T}(p)$. Under favourable hypotheses, the degree introduced in Section 7.3 can be used to prove the existence of periodic solutions. A similar idea is used in the following result, due to Lazer (1968); the mapping corresponding to $\mathscr{T}$ is single-valued.

Theorem 9.2.11 *Consider the equation*

$$\dot{x} = f(x) + F(x, t) \quad (x \in \mathbf{R}^n), \tag{9.2.7}$$

where f is C^1, and F is bounded, continuous and of period ω in t. Suppose that whenever $p \in \mathscr{B}$, the equation

$$\dot{x} = f(x) + F(p(t), t)$$

has a unique ω-periodic solution $\mathscr{T}(p)$, and that $\|\mathscr{T}(p)\| < A$, a constant independent of p. Then (9.2.7) *has a solution of period ω.*

Proof It is shown that $\mathscr{T}$ is a compact mapping on the ball $B(0, A)$ – we omit the details. Schauder's fixed point theorem yields the existence of a function $p_0 \in \mathscr{B}$ satisfying (9.2.7).

Another way in which Leray–Schauder degree is used can be illustrated by a consideration of the third order equation

$$\dddot{x} + \psi(\dot{x})\ddot{x} + \phi(x)\dot{x} + \theta(x) = p(t) \quad (x \in \mathbf{R}) \tag{9.2.8}$$

(see Reissig (1972*b*), Ezeilo (1975)). This method is useful for equations of this type when simpler techniques fail. We adopt the common convention that capital letters denote indefinite integrals: $P(t) = \int_0^t p(\tau)\,d\tau$, $\Phi(y) = \int_0^y \phi(\xi)\,d\xi$, and so on. The hypotheses relating to (9.2.8) are the following (K, a, b, m are constants): (i) ψ, ϕ, θ, p are continuous functions of their arguments, and p is of period ω in t;(ii)$|P(t)| \leqslant m$ (so that p has mean value zero), (iii) $\theta(y)$ sign $y \geqslant 0$ ($|y| \geqslant a$), (iv) $|\Phi(y) - by| \leqslant M$ ($b \leqslant 0$). Note that there is no condition on ψ (other than continuity).

Equation (9.2.8) is embedded in a system depending on a parameter $\mu \in [0, 1]$, and involving arbitrary positive constants γ, K:

$$\dot{x} = Ax + f(x, t, \mu) \quad (x = (x_1, x_2, x_3)), \tag{9.2.9}$$

where

$$A = \begin{pmatrix} 0 & 1 & 0 \\ 0 & 0 & 1 \\ -\gamma & -b & 0 \end{pmatrix}, f = \begin{pmatrix} 0 \\ bx_1 - \Phi_\mu(x_1) - \Psi_\mu(x_2) + \mu P(t) \\ \gamma x_1 - \theta_\mu(x_1) \end{pmatrix}.$$

Here
$$(2 - \mu)\Phi_\mu(x_1) = \Phi(x_1) - (1 - \mu)\Phi(-x_1),$$

$$(2-\mu)\Psi_\mu(x_2) = \Psi(x_2) - (1-\mu)\Psi(-x_2),$$

and $$\theta_\mu(x_1) = \mu\theta(x_1) + (1-\mu)Kx_1/(1+|x_1|).$$

The value $\mu = 1$ of the parameter corresponds to the given equation. For $\mu = 0$, the right hand side of (9.2.9) is an odd function of x. Referring back to (9.1.5), the periodic solutions of (9.2.9) are solutions of

$$(I - e^{-\omega A})x(t) = -\int_t^{t+\omega} e^{(t-s)A} f(x(s), s, \mu)\, ds.$$

Now as long as $\gamma \neq 0, I - e^{-\omega A}$ is a non-singular matrix, so this equation can be written as an operator equation in the space $\mathscr{B}$:

$$V_\mu(x) \equiv x - T_\mu(x) = 0, \tag{9.2.10}$$

where $T_\mu(x)$ is the function

$$\int_t^{t+\omega} (e^{-\omega A} - I)^{-1} e^{(t-s)A} f(x(s), s, \mu)\, ds.$$

The reader can easily check that T_μ is a compact mapping. Since f is an odd function of x when $\mu = 0$, V_0 is an odd mapping. Thus, if B_r is the ball $B(0, r)$ in $\mathscr{B}$, $d(V_0, B_r, 0)$ is odd if it is defined (Theorem 4.3.11). Suppose that an *a priori* bound can be found for the solutions of (9.2.10) – that is, all zeros of V_μ are known to be contained in B_R, say, where R is independent of μ. Then, by the invariance under homotopy of degree, $d(V_1, B_R, 0) = d(V_0, B_R, 0)$, which is non-zero; hence there is a periodic solution χ of (9.2.8) with $|\chi(t)| < R$ (all t).

Theorem 9.2.12 *Under the given hypotheses, there exists a periodic solution of* (9.2.8) *of period* ω.

Proof We obtain a bound for the possible periodic solutions of (9.2.9); in the light of the above remarks, the theorem will then have been proved. The system (9.2.9) corresponds to a third order equation for $x \equiv x_1$:

$$\dddot{x} + \psi_\mu(\dot{x})\ddot{x} + \phi_\mu(x)\dot{x} + \theta_\mu(x) = \mu p(t). \tag{9.2.11}$$

Multiplying this equation by $\dot{x}$ and integrating, we have

$$[\ddot{x}\dot{x} + \int_0^{\dot{x}} y\psi_\mu(y)\,dy + \dot{x}\int_0^x \phi_\mu(y)\,dy + \int_0^x \theta_\mu(y)\,dy - \mu\dot{x}P(t)]_0^\omega$$
$$= \int_0^\omega \ddot{x}(\ddot{x} + \Phi_\mu(x) - \mu P(t))\,dt.$$

From now on we suppose that $x(t)$ is a periodic solution; all integrals will be over $[0,\omega]$. The last equation reduces to

$$\int \ddot{x}(\ddot{x} + \Phi_\mu(x) - \mu P(t))\,dt = 0.$$

This can be written

$$\int \ddot{x}^2 = \int b\dot{x}^2 - \int \ddot{x}[\Phi_\mu(x) - bx - \mu P(t)]\,dt,$$

whence, using conditions (ii) and (iv),

$$\int \ddot{x}^2 \leqslant \int |\ddot{x}|(M+m).$$

But $\int|\ddot{x}| \leqslant (\omega\int\ddot{x}^2)^{1/2}$, so that

$$\int \ddot{x}^2 \leqslant \omega(M+m)^2.$$

Since $x(t)$ is periodic, $\dot{x}$ has a zero in $[0,\omega]$: say $\dot{x}(\tau)=0$. Then $\dot{x}(t) = \int_\tau^t \ddot{x}(s)ds$ $(\tau \leqslant t \leqslant \tau+\omega)$, whence

$$|\dot{x}(t)| \leqslant \omega(M+m). \tag{9.2.12}$$

Integrating (9.2.11) gives us $\int\theta_\mu(x)\,dt = 0$. Thus, by (iii), there is τ_1 such that $|x(\tau_1)| < a$. It follows that

$$|x(t)| < a + \omega^2(M+m) \quad (0 \leqslant t \leqslant \omega). \tag{9.2.13}$$

We see from (9.2.9) that $\dot{x}_3$ depends only on x_1 and x_2. The estimates (9.2.12) and (9.2.13) are therefore enough to imply the existence of R, independent of μ, such that every periodic solution $x(t)$ of (9.2.9) satisfies $\|x\| < R$. The proof is complete.

Remark There are adaptations of the theorem to cover other hypotheses on the functions ψ, ϕ, θ – for example, $\theta(x)\operatorname{sign} x < 0$ or $\theta \equiv \theta(x, \dot{x}, \ddot{x}, t)$ (in the latter case a condition on ψ is required).

9.3 Holomorphic mappings and differential equations

The degree $d(\phi, D, p)$ of a mapping ϕ is an algebraic count of the number of its p-points in the set D. If the ambient space is $\mathbf{C}^n$ (that is $\mathbf{R}^{2n}$ endowed with complex structure), and ϕ is holomorphic, we shall show that an index $i(\phi, \zeta, p)$ is necessarily non-negative; thus $d(\phi, D, p)$ is in these circumstances an exact count of the number of p-points of ϕ in D. This section is devoted to proving this and exploring some of its consequences in the theory of differential equations.

To fix our ideas: we work in $\mathbf{C}^n$, the space of n-tuples $z = (z_1, \ldots, z_n)$ of complex numbers; D is a bounded, open subset of $\mathbf{C}^n$; $p \in \mathbf{C}^n$; $\phi \in \mathscr{H}(\bar{D})$, the vector space of holomorphic mappings of $\bar{D}$ into $\mathbf{C}^n$; we write $|z| = \max\{|z_1|, \ldots, |z_n|\}$. We can view ϕ also as a mapping from a subset of $\mathbf{R}^{2n}$ into $\mathbf{R}^{2n}$; in this way $d(\phi, D, p)$ is defined (provided $p \notin \phi(\partial D)$).

Theorem 9.3.1 *Suppose that D is an open, bounded subset of $\mathbf{C}^n$, and $\phi \in \mathscr{H}(\bar{D})$. If $p \notin \phi(\partial D)$, then $d(\phi, D, p) \geqslant 0$.*

Proof We show, essentially, that a holomorphic mapping is orientation-preserving; this is the geometric interpretation of the theorem. It is sufficient to give a proof under the hypothesis that p is not a critical value of ϕ; the general result is then obtained by the usual approximation argument. Let $z_j = x_j + iy_j$ and $\phi_j = u_j + iv_j (j = 1, \ldots, n)$. Each ϕ_j is a holomorphic function of n complex variables, so the Cauchy–Riemann equations are satisfied:

$$\frac{\partial u_k}{\partial x_j} = \frac{\partial v_k}{\partial y_j}, \quad \frac{\partial u_k}{\partial y_j} = -\frac{\partial v_k}{\partial x_j} \quad (k, j = 1, \ldots, n).$$

Consider ζ such that $\phi(\zeta) = p$. The matrix of $\phi'(\zeta)$ regarded as a mapping of $\mathbf{C}^n$ into itself is (with respect to the standard basis of $\mathbf{C}^n$) $A = (\alpha_{kj})$, where

$$\alpha_{kj} = \frac{\partial u_k}{\partial x_j} - i\frac{\partial u_k}{\partial y_j}.$$

By considering the Jordan canonical form of A, it can be checked that $\det(\phi'(\zeta)) \geqslant 0$; since we are supposing that $\phi'(\zeta)$ is non-singular, we have that $i(\phi, \zeta, p) = +1$. If p is critical, the usual approximation process tells us that $i(\phi, \zeta, p) \geqslant 0$ (but not yet that $i(\phi, \zeta, p) \geqslant 1$). The result now follows easily.

Definition 9.3.2 Suppose that ϕ is a holomorphic map and ζ an isolated solution of $\phi(z) = p$. The *multiplicity* of ζ as a p-point of ϕ is $i(\phi, \zeta, p)$; ϕ is *simple* if its multiplicity is 1.

Note that this definition is consistent with Definition 9.2.5. Next, we have the following result (Lloyd, 1975).

Theorem 9.3.3 *The multiplicity of a point ζ as a p-point of ϕ is at least* 1; *it is greater than* 1 *if and only if ζ is a critical point of ϕ.*

Proof We saw in the proof of Theorem 9.3.1 that $i(\phi, \zeta, p) = 1$ if ζ is not a critical point of ϕ. If ζ is a critical point, take U to be a neighbourhood of ζ such that $i(\phi, \zeta, p) = d(\phi, U, p)$. Now $V = \phi(U)$ is a neighbourhood of p (see, for example, Hervé (1963), Chapter 4); choose $p_1 \in V$ to be a non-critical value of ϕ (possible by Sard's theorem) with $d(\phi, U, p_1) = i(\phi, \zeta, p)$. Now U contains p_1-points of ϕ, at each of which the index of ϕ is, of course, $+1$. It follows that $i(\phi, \zeta, p) \geqslant 1$.

For the second part of the theorem, we take, without loss of generality, $\zeta = p = 0$. The matrix $\phi'(0)$ has at least one zero eigenvalue when 0 is a critical point. By a suitable change of co-ordinates, we may suppose $\phi'(0)$ to be in Jordan canonical form, whence ϕ is, in the new co-ordinates, given by $(\zeta_1, \ldots, \zeta_n) \mapsto (h_1, \ldots, h_n)$, where the h_i are power series in $\zeta_1, \ldots, \zeta_n$, and h_1 has no linear part. By the calculations that appear in Section 9 of Chapter I of Cronin (1964), it is seen that the multiplicity is greater than 1.

We can now formulate an analogue of Rouché's theorem.

Theorem 9.3.4 *Let D be a bounded, open set in $\mathbf{C}^n$; suppose that $f, g \in \mathscr{H}(\bar{D})$ are such that $|g(z)| < |f(z)|$ if $z \in \partial D$. Then f has finitely many zeros in D, and, counting multiplicity, f and $f + g$ have the same number of zeros in D.*

Proof Since $|g| < |f|$ on ∂D, neither f nor $f+g$ has a zero in ∂D. Thus $d(f, D, 0)$ and $d(f+g, D, 0)$ are both defined. Let

$$H(\lambda, z) = f(z) + \lambda g(z) \quad (z \in \bar{D}, 0 \leqslant \lambda \leqslant 1).$$

If $H(\lambda, z) = 0$ with $z \in \partial D$ and $\lambda \neq 0$, we have $|f(z)| = \lambda|g(z)| < \lambda|f(z)| \leqslant |f(z)|$, which is absurd. By invariance under homotopy, therefore, $d(f, D, 0) = d(f+g, D, 0)$. Now a compact analytic set in $\mathbf{C}^n$ is finite (see Remmert and Stein (1953)). Since f and $f+g$ are non-zero on ∂D, both $D \cap f^{-1}(0)$ and $D \cap (f+g)^{-1}(0)$ are compact analytic sets. Let the distinct zeros of f in D be $s_1, \ldots, s_r$. Then

$$d(f, D, 0) = \sum_{i=1}^{r} i(f, s_i, 0).$$

So, provided account is taken of multiplicity, $d(f, D, 0)$ is precisely the number of zeros of f in D. Together with the corresponding statement about $f+g$, this gives the result.

We now go on to use Theorem 9.3.4 in the consideration of the periodic solutions of differential equations of the form

$$\dot{z} = f(z, t) \quad (z \in \mathbf{C}^n, t \in \mathbf{R}), \tag{9.3.1}$$

where f is everywhere holomorphic in z, and is continuous in t and of period ω. As usual, we identify the equation (9.3.1) with the function f. Let $\mathscr{S}$ be the collection of such equations, endowed with the Fréchet topology generated by the seminorms

$$p_k(f) = \max\{|f_i(z, t)|; 1 \leqslant i \leqslant n, 0 \leqslant t \leqslant \omega, |z| \leqslant k\} \quad (k = 1, 2, \ldots).$$

Recall that a Fréchet space is a complete, metrisable, locally convex topological vector space. Let σ be a metric on $\mathscr{S}$ generating the topology. The p_k are, in fact, norms in our case.

Theorem 9.3.5 *Suppose that $f \in \mathscr{S}$ has exactly M solutions of period ω. There is a neighbourhood $U(f)$ of f in $\mathscr{S}$ such that, if $g \in U(f)$, then g has at least M solutions of period ω. Near a solution of f of multiplicity k, g has exactly k ω-periodic solutions.*

Proof Let the ω-periodic solutions of f be $\psi_1(t), \ldots, \psi_r(t)$, where ψ_i has multiplicity k_i. Following the ideas of Section 9.1, define the function $q: \mathscr{S} \times \mathbf{C}^n \to \mathbf{C}^n$ by

$$q(f,c) = z_f(\omega;0,c) - c.$$

Writing $\psi_i(0) = \xi_i$, we have $q(f,\xi_i) = 0$ $(i = 1,\ldots,r)$ and $q(f,c) \neq 0$ if $c \neq \xi_1,\ldots,\xi_r$. Choose disjoint closed balls $B_i = \overline{B(\xi_i,\alpha_i)}$ $(i = 1,\ldots,r)$ so that $q(f,.)$ is defined and holomorphic in each B_i. Then choose J so that $\bigcup_{i=1}^{r} B_i \subset B(0,J)$. There is $\varepsilon_0 > 0$ such that $q(g,.)$ is defined in each B_i if $p_J(f-g) < \varepsilon_0$. Theorem 9.3.4 now tells us that, if $\varepsilon < \varepsilon_0$ is small enough, $q(g,.)$ has exactly k_i zeros in $B_i (i = 1,\ldots,r)$ for every equation g satisfying $p_J(f-g) < \varepsilon$. Thus such g have at least $\Sigma k_i = M$ solutions of period ω. The proof is complete.

We can use the last result to prove that a periodic solution of an equation such as (9.3.1), whose multiplicity is $k > 1$, can be separated into k simple (that is, multiplicity 1) solutions by a suitable small change in the equation. This result should be compared with Theorem 9.2.4.

Theorem 9.3.6 *Let $\psi(t)$ be a solution of $f \in \mathscr{S}$ of period ω, whose multiplicity is $k > 1$. Given $\delta > 0$, there is $g \in \mathscr{S}$ with $\sigma(f,g) < \delta$ such that g has k simple ω-periodic solutions $\chi_1,\ldots,\chi_k$ with $|\chi_i(0) - \psi(0)| < \delta$ $(i = 1,\ldots,k)$.*

Proof Let ε be real, and define

$$g_\varepsilon(z,t) = f(z,t) - \varepsilon(z - \psi(t)). \tag{9.3.2}$$

Clearly ψ is an ω-periodic solution of g_ε; the variational equation (see (9.2.4)) is

$$\dot{\zeta} = (A(t) - \varepsilon I)\zeta, \tag{9.3.3}$$

where I is the identity matrix and $A(t) = (\partial f_i/\partial z_j)(\psi(t),t)$. The solutions of (9.3.3) are given by $a(t) = e^{-\varepsilon t}b(t)$, where $b(t)$ is the solution of $\dot{\zeta} = A(t)\zeta$ with $b(0) = a(0)$. So choose κ that $e^{\kappa\omega}$ is the smallest of the real positive characteristic multipliers of the equation $\dot{\zeta} = A(t)\zeta$ which are greater than 1 (take $\kappa = \infty$ if no such multipliers exist). Then, if $0 < \varepsilon < \kappa$, (9.3.3) has no solutions of period ω, so that, by Theorem 9.2.7, ψ is a simple ω-periodic solution of g_ε. Now choose $\Delta < \delta/k$ to be such that $\psi(t)$ is the only ω-periodic solution of f whose initial point is in the ball $B(\psi(0),\Delta)$. Theorem 9.3.5 enables us to choose $\varepsilon_1 < \kappa$ so that $\sigma(f,g_{\varepsilon_1}) < \Delta$ and $B(\psi(0),\Delta)$ contains the initial points of k and only k solutions of $g_{\varepsilon_1} \equiv g_1$ of period ω (counting multiplicity). If these are $\psi, u_1, \ldots, u_s$, let $\eta = \min\{|\psi(0) - u_i(0)|; i = 1, \ldots, s\}$. Repeating the procedure just described, there is g_2 with $\sigma(g_1, g_2) < \eta < \Delta$ with u_1 as a simple ω-periodic solution. When $\sigma(g_1, g_2)$ is sufficiently small, the solutions $\psi, u_1, \ldots, u_s$ of g_1 at $t = 0$ are shifted by at most $\frac{1}{2}\eta$; thus, by Theorem 9.3.5, g_2 has, in addition to u_1, a simple ω-periodic solution with initial point in $B(\psi(0), \frac{1}{2}\eta)$. Proceeding in this way, we arrive after at most k steps at an equation g with $\sigma(f,g) < \delta$ having k simple ω-periodic solutions with initial points in $B(\psi(0), \delta)$.

Remarks (1) Theorems 9.3.5 and 9.3.6 have extensions covering solutions of (9.3.1) of period $m\omega$ $(m > 1)$.

(2) It is not generally possible to conclude that small changes in the equation (9.3.1) leave the number of periodic solutions unchanged. For this to be true, a certain set of equations $\mathscr{S}_1$ has to be excluded; $\mathscr{S}_1$ consists of those equations with so-called singular periodic solutions. Under a certain additional hypothesis, equations in the same component of $\mathscr{S} \setminus \mathscr{S}_1$ have the same number of periodic solutions. For the details the reader is referred to Lloyd (1975).

It is clear from the preceding that much more information may be obtained about the precise number of solutions of an equation $\phi(x) = p$ when ϕ is a holomorphic map and x is a complex vector than when x is real. This leads to a useful way of bounding the number of zeros in the real case – simply consider the complexification of ϕ. Results along these lines have been obtained, in varying degrees of generality, by Cronin (1971), Nussbaum (1971*b*), Schwartz (1963), Thomas (1974). In line with our policy hitherto, we consider the simplest case – the Brouwer degree. Again D is a bounded, open subset of $\mathbf{R}^n$, $p \in \mathbf{R}^n$ and $\phi: \bar{D} \to \mathbf{R}^n$ is continuous.

Lemma 9.3.7 *Suppose that $p \notin \phi(\partial D)$. If U is a sufficiently small neighbourhood of p, there is a set N, open and dense in U, such that if $q \in N$, then (a) $\phi^{-1}(q)$ is finite and (b) the number of points m_q of $\phi^{-1}(q)$ satisfies*

$$|d(\phi, D, p)| \leqslant m_q; \quad d(\phi, D, p) \equiv m_q \pmod 2.$$

Proof Let U be so small that $d(\phi, D, p) = d(\phi, D, q)$ if $q \in U$. The existence of a set N, open and dense in U, satisfying (a) is assured by Sard's theorem. Let $\phi^{-1}(q) = \{y_1, \ldots, y_m\}$; each of these q-points is simple – that is, $i(\phi, y_i, q) = \pm 1$. The result follows from the relation

$$d(\phi, D, p) = \sum_{i=1}^{m} i(\phi, y_i, q).$$

With the same notation as the lemma, we obtain an upper bound for m_q by considering a complexification ϕ^* of ϕ. We write $x^* = (x,0) \in \mathbf{C}^n$ if $x \in \mathbf{R}^n$ and $\bar{z}$ for the complex conjugate of z. In the following $D^* = D \times D_1$, where D_1 is to be a symmetric neighbourhood of 0 in $\mathbf{R}^n$.

Theorem 9.3.8 *Suppose that D^* and $\phi^* \in \mathscr{H}(\overline{D^*})$ can be found such that* (a) $\phi^* = \phi$ *on* D, (b) $\phi^*(\bar{z}) = \overline{\phi^*(z)}$ ($z \in D^*$), (c) *if q is not a critical value of ϕ, neither is q^* a critical value of ϕ^*, and* (d) $d(\phi^*, D^*, p^*)$ *is defined. There is a neighbourhood U of p in $\mathbf{R}^n$ and an open, dense subset N of U such that, if $q \in N$, the number m_q of q-points of ϕ satisfies*

$$|d(\phi, D, p)| \leqslant m_q \leqslant d(\phi^*, D^*, p^*) \tag{9.3.4}$$

and

$$d(\phi, D, p) \equiv m_q \equiv d(\phi^*, D^*, p^*) \pmod 2. \tag{9.3.5}$$

Proof Since $d(\phi^*, D^*, p^*)$ is defined, so is $d(\phi, D, p)$. The left hand parts of (9.3.4) and (9.3.5) were proved in Lemma 9.3.7. Take U and N as in the lemma; N was so chosen that every $q \in N$ is not a critical value of ϕ and $d(\phi, D, p) = d(\phi, D, q)$. Suppose further that U is so small that $d(\phi^*, D^*, p^*) = d(\phi^*, D^*, q^*)$ if $q \in U$. If $q \in N$, q^* is not a critical value of ϕ^*, by hypothesis. Thus each q^*-point of ϕ^* in D^* has index $+1$ (Theorem 9.3.1). But $m_q \leqslant$ number of q^*-points of $\phi^* = d(\phi^*, D^*, q^*) = d(\phi^*, D^*, p^*)$. Thus (9.3.4) is proved. To prove (9.3.5), we observe that the only q^*-points of ϕ^* in a sufficiently small neighbourhood F of $D \times \{0\}$ are the q-points of ϕ in D – for the q^*-points of ϕ^* are isolated in D^*. Therefore

$$d(\phi^*, D^*, q^*) = d(\phi^*, D^* \setminus \bar{F}, q^*) + m_q.$$

Now if $\phi^*(\zeta) = q^*$ with $\zeta \in D^*\backslash D$, then $\phi^*(\bar{\zeta}) = q^*$ (and $\bar{\zeta} \in D^*\backslash D$); it follows that $d(\phi^*, D^*\backslash \bar{F}, q^*)$ is even– it is, after all, the number of q^*-points of ϕ^* in $D^*\backslash D$. This completes the proof.

Remark The hypotheses of Theorem 9.3.8 are comparatively mild, and are easily shown to hold in many cases. D^* is chosen to optimise the estimates (9.3.4) and (9.3.5).

The proofs of Lemma 9.3.7 and Theorem 9.3.8 depend on two important facts – Sard's theorem, and the positivity of the degree of holomorphic mappings. Both these can be extended to cover mappings defined on infinite dimensional spaces. Using Smale's version of Sard's theorem (Theorem 8.2.1) and a result of Nussbaum (1971*b*), the result of Theorem 9.3.8 holds when ϕ is defined on a bounded, open subset of a Banach space, satisfies the conditions used in Section 8.2, and is analytic. The degree that is used is then that defined in Definition 8.2.2. In particular the estimates (9.3.4) and (9.3.5) hold when $\phi = I - f$ and f is compact or a strict set contraction. In the latter case, if ϕ is also analytic, it is not very difficult to see that the positivity of the degree follows by an approximation argument, first approximating to a strict set contraction by a suitable compact mapping, and then approximating to a compact mapping by one of finite dimensional range.

9.4 Boundary value problems

Several authors have studied boundary value problems of the form

$$\left.\begin{aligned} \dot{x} &= f(x, t) \\ x(\omega) &= x(0) \end{aligned}\right\} \quad (x \in \mathbf{R}^m), \tag{9.4.1}$$

where f is, of course, no longer required to be periodic. We mention a few examples: Mawhin (1970), Knobloch (1971), Schmitt (1972), Bebernes and Schmitt (1973), Bebernes (1974). Further references may be found in these papers.

We illustrate the use of degree theory in these problems by considering an equation

$$x^{(n)} = f(x, \dot{x}, \ldots, x^{(n-1)}, t), \tag{9.4.2}$$

where $x \in \mathbf{R}^m$ and $f: \mathbf{R}^{nm} \times [0,1] \to \mathbf{R}^m$ is continuously differentiable. Writing $y = (\xi_1, \ldots, \xi_n) \in \mathbf{R}^{nm}$, where ξ_i are themselves m-vectors, (9.4.2) is equivalent to a system

$$\dot{y} = F(y,t) \quad (y \in \mathbf{R}^{nm}), \tag{9.4.3}$$

where $F(y,t) = (\xi_2, \ldots, \xi_n, f(\xi_1, \ldots, \xi_n, t))$. Solutions of (9.4.2) satisfying the boundary conditions

$$x^{(i)}(0) = x^{(i)}(\omega) \quad (i = 0, \ldots, n-1) \tag{9.4.4}$$

are sought, where $\omega \in (0,1)$ but is not at this stage specified. The idea of ω-non-recurrence introduced in Section 9.2 is again useful. In terms of y, condition (9.4.4) is just $y(0) = y(\omega)$, so that we seek zeros of the mapping $c \mapsto y_F(\omega; 0, c) - c$ $(c \in \mathbf{R}^{nm})$.

Theorem 9.4.1 (Schmitt, 1972) *Let D be an open, bounded subset of $\mathbf{R}^m$, and let $h: \mathbf{R}^m \to \mathbf{R}^m$ be defined by*

$$h(x) = f(x, 0, \ldots, 0).$$

Suppose that the Brouwer degree $d(h, D, 0)$ is defined and non-zero. Then there is $\omega_0 > 0$ such that (9.4.2) *has a solution satisfying* (9.4.4) *for each $\omega \in (0, \omega_0)$.*

Proof (i) Take B to be a bounded, open set containing $\bar{D}$. Given $A > 0$, there is certainly an $\omega_1 \in (0,1)$ such that any solution $\phi(t)$ of (9.4.2) with $\phi(0) \in \bar{D}$ and $|\phi^{(i)}(0)| \leqslant A \, (i = 1, \ldots, n-1)$ satisfies

$$\phi(t) \in \bar{B}, \quad |\phi^{(i)}(t)| \leqslant 2A \quad (i = 1, \ldots, n-1; 0 \leqslant t \leqslant \omega_1).$$

Define $\quad \Omega = \{(\xi_1, \ldots, \xi_n); \xi_1 \in D, |\xi_i| < A \, (i = 2, \ldots, n)\}$

and $\quad \Sigma = \{(\xi_1, \ldots, \xi_n); \xi_1 \in B, |\xi_i| < 2A \, (i = 2, \ldots, n)\}$;

Ω and Σ are subsets of $\mathbf{R}^m \times \ldots \times \mathbf{R}^m$ (n copies). For $K \subset \bar{\Omega}$, let $S(K)$ be the tube of solutions of (9.4.3) for $0 \leqslant t \leqslant \omega_1$ with initial points in K. It is seen that $S(K)$ is always a relatively compact subset of $C([0, \omega_1], \mathbf{R}^{nm})$, the space of continuous functions from $[0, \omega_1]$ into $\mathbf{R}^{nm}$ with the supremum norm.

(ii) The next stage is to show that $\partial\Omega$ consists entirely of points of ω-non-recurrence from $t = 0$ if ω is sufficiently small. If this is not so, there are sequences (t_n) and (η_n) in $\mathbf{R}$ and $\partial\Omega$, respectively, with $t_n \to 0$ and $y_F(t_n;0,\eta_n)=\eta_n$. By the last sentence of (i), there is a subsequence of (η_n), which we take to be the whole sequence, such that $\eta_n \to \eta_0 \in \partial\Omega$ and $\phi_n(t)=y_F(t;0,\eta_n) \to y_F(t;0,\eta_0)$ uniformly. Now

$$\phi_n(t)=\eta_n+\int_0^t F(\phi_n(\tau),\tau)\,d\tau,$$

so that $t_n^{-1}\int_0^{t_n} F(\phi_n(\tau),\tau)\,d\tau=0$. By the continuity of F, it follows that $F(\eta_0,0)=0$, and hence that there is $\zeta \in \partial D$ with $h(\zeta)=0$. This is a contradiction, so the statement at the beginning of (ii) is proved for $\omega<\omega_0$, say.

(iii) Take Ω and ω_0 as defined in (i) and (ii). Consider the homotopy

$$H(\lambda,y)=(\xi_2,\ldots,\xi_n,f(\xi_1,\lambda\xi_2,\ldots,\lambda\xi_n,0)) \quad (0\leqslant\lambda\leqslant 1,\ y\in\bar{\Omega}).$$

If $H(\lambda,y)=0$ with $y\in\partial\Omega$, then $\xi_2=\ldots=\xi_n=0$ and $h(\xi_1)=0$. But then $\xi_1\in\partial D$, and we have $0\in h(\partial D)$, contrary to hypothesis. By homotopy invariance,

$$d(F(.,0),\Omega,0)=d(G,\Omega,0),$$

where $G(y)=(\xi_2,\ldots,\xi_n,h(\xi_1))$. By reasoning similar to that of the proof of Lemma 4.2.3, $|d(G,\Omega,0)|=|d(h,D,0)|$. From this and the hypothesis $d(h,D,0)\neq 0$, it follows that $d(F(.,0),\Omega,0)\neq 0$.

(iv) By (ii), (iii) and Theorem 9.2.2, we deduce that (9.4.2) has a solution satisfying (9.4.4) for $\omega<\omega_0$.

Corollary 9.4.2 *The theorem applies if $m>1$ and the inner product $x\cdot h(x)$ is non-zero on some sphere S centre the origin.*

Proof Since $m>1$, S is connected. If $x\cdot h(x)\neq 0$ on S, then $h(x)$ and $h(-x)$ are never in the same direction for $x\in S$. It follows from the odd mapping theorem (Theorem 3.2.6) that $d(h,D,0)$ is odd, and therefore non-zero.

From the theorem, we can derive the following result for the Liénard type equation

$$\ddot{x}+k(x,\dot{x})\dot{x}+g(x)=p(t) \quad (x\in\mathbf{R}). \tag{9.4.5}$$

Corollary 9.4.3 *Suppose that there are $x_1 > 0$ and $x_2 < 0$ such that $g(x_1) - p(0)$ and $g(x_2) - p(0)$ are of opposite signs. Then there is a solution of* (9.4.5) *satisfying* (9.4.4) *if ω is sufficiently small.*

Proof In the notation of the theorem, $h(x) = p(0) - g(x)$. Let $D = (x_2, x_1)$. Then $h(x_2)$ and $h(x_1)$ have opposite signs, whence $d(h, D, 0) \neq 0$, by the odd mapping theorem.

Remarks (1) Corollary 9.4.3 involves no condition on the damping coefficient k.

(2) We can also look at (9.4.5) when $x \in \mathbf{R}^m$ with $m > 1$. The result is valid if, for example, $x \cdot (g(x) - p(0))$ is non-zero on some sphere centre 0 (by Corollary 9.4.2).

Another version of problem (9.4.1) is to seek solutions when ω is fixed. For example, Bebernes (1974) considered the following problem for a second order system:

$$\left.\begin{aligned} \ddot{x} &= f(x, \dot{x}, t) \\ x(0) = x(1), &\quad \dot{x}(0) = \dot{x}(1) \end{aligned}\right\} \quad (0 \leqslant t \leqslant 1, x \in \mathbf{R}^n), \tag{9.4.6}$$

where f is supposed to be C^1. Problem (9.4.6) is 'embedded' in the family

$$\left.\begin{aligned} &\ddot{x} = \lambda f(x, \dot{x}, t) + (1 - \lambda)x \quad (0 \leqslant \lambda \leqslant 1) \\ &x(0) = x(1), \quad \dot{x}(0) = \dot{x}(1). \end{aligned}\right\} \tag{9.4.7}$$

Problem (9.4.7) is converted to an operator equation $x = \lambda T(x)$, where T is a compact mapping defined on the space $\mathscr{B}$ of C^1 functions satisfying the boundary conditions of (9.4.7). Suppose now that D is a bounded, open subset of $\mathbf{R}^n \times \mathbf{R}^n$ with the property that a solution of (9.4.7) meeting ∂D must also meet $\mathscr{C}D$. This means that there is no solution of (9.4.7) in ∂W, where

$$W = \{x \in \mathscr{B}; (x(t), \dot{x}(t)) \in D \, (0 \leqslant t \leqslant 1)\}.$$

Hence the Leray–Schauder degree $d(I - \lambda T, W, 0)$ is defined for $\lambda \in [0, 1]$, and is independent of λ. Consequently $d(I - T, W, 0) = 1$, and there is a solution of (9.4.6) entirely contained in D.

9.5 Bifurcation theory

Many problems in applied mathematics require, for their resolution, knowledge of the solutions of operator equations of the form

$$u = G(\lambda, u), \tag{9.5.1}$$

where $\lambda \in \mathbf{R}$ and $u \in \mathscr{B}$, a real Banach space. Such equations are related to eigenvalue problems. We suppose that G is a compact mapping of $\mathscr{E} = \mathbf{R} \times \mathscr{B}$ into $\mathscr{B}$ and is of the form

$$G(\lambda, u) = \lambda L(u) + H(\lambda, u), \tag{9.5.2}$$

where L is linear and compact, and $H(\lambda, u) = O(\| u \|^2)$ uniformly on compact λ-intervals as $u \to 0$. The norm on $\mathscr{E}$ which we use is

$$\| (\lambda, u) \| = (|\lambda|^2 + \| u \|^2)^{1/2}.$$

A solution of (9.5.1) is a pair $(\lambda, u) \in \mathscr{E}$. Clearly $(\lambda, 0)$ is always a solution; the curve $\mathscr{C}_0 = \{(\lambda, 0); \lambda \in \mathbf{R}\}$ is the curve of so-called *trivial solutions*. We write $\mathscr{S}_0$ for the set of non-trivial solutions of (9.5.1) and $\mathscr{S} = \overline{\mathscr{S}_0}$.

Definition 9.5.1 Suppose that $u(t)$ is a simple curve $\mathscr{C}$ of solutions of (9.5.1). If there is a solution in every neighbourhood of $u(\tau)$ which is not contained in $\mathscr{C}$, then $u(\tau)$ is said to be a *bifurcation point* of G with respect to the curve $\mathscr{C}$.

We shall, in the main, study the bifurcation points of G with respect to the curve $\mathscr{C}_0$ of trivial solutions. Such points correspond to values of λ in every deleted neighbourhood of which (9.5.1) has solutions of 'small norm'.

Our basic references are Sattinger (1971, 1973) and Rabinowitz (1971). More general results to which we briefly refer may be found in Crandall and Rabinowitz (1971, 1973), and McLeod and Sattinger (1973). The compactness assumption on G can be relaxed – see Stuart (1973) and Dancer (1975).

The term 'bifurcation' is one that is also used in topological dynamics and catastrophe theory. There, a family of functions or vector fields depending on a parameter λ is considered (λ may be a vector). The bifurcation values of λ are those where the system is not structurally stable – that is, where a small change in λ changes the topological nature of the collection of orbits. We do not pursue these problems.

We write $r(L) = \{\mu ; \mu^{-1}$ is a real eigenvalue of $L\}$; the members of $r(L)$ are the characteristic values of L. We shall sometimes say that 'λ is a bifurcation value' rather than '$(\lambda, 0)$ is a bifurcation point'.

Theorem 9.5.2 *The bifurcation points of G with respect to $\mathscr{C}_0$ are contained in the set $\{(\lambda, 0); \lambda \in r(L)\}$.*

Proof Suppose that $\lambda_0 \notin r(L)$. Then $I - \lambda_0 L$ is invertible, so that there is $k > 0$ such that $\|(I - \lambda_0 L)(u)\| \geqslant k\|u\|$. We have, for $u \neq 0$,

$$\begin{aligned}\|u - G(\lambda, u)\| &\geqslant \|(I - \lambda_0 L)(u)\| - |\lambda - \lambda_0| \, \|L(u)\| - \|H(\lambda, u)\| \\ &> k\|u\| - |\lambda - \lambda_0| \, \|L\| \, \|u\| + O(\|u\|^2) \\ &> 0\end{aligned}$$

if $|\lambda - \lambda_0|$ and $\|u\|$ are sufficiently small. So under these conditions $u = G(\lambda, u)$ has the unique solution $u = 0$, and sufficiently small neighbourhoods of $(\lambda_0, 0)$ in $\mathscr{E}$ contain no non-trivial solutions.

It is certainly not true that every characteristic value of L is a bifurcation value of G; indeed, the basic problem is to characterise those characteristic values at which bifurcation occurs. Before proceeding, recall that the *multiplicity* of $\mu \in r(L)$ is

$$\dim \bigcup_{j=1}^{\infty} \ker \{(\mu L - I)^j\};$$

since L is compact, this is finite (see Grothendieck (1973), Chapter 5, say).

Lemma 9.5.3 *Suppose that, for $\lambda = \lambda_0$, there is $\varepsilon > 0$ such that* (9.5.1) *has no solution of norm ε in $\mathscr{B}$. Then there is $\delta > 0$ such that* (9.5.1) *has no solutions of norm ε whenever $|\lambda - \lambda_0| < \delta$.*

Proof If the conclusion is false, there is a sequence $\{(\lambda_n, u_n)\}$ with $u_n = G(\lambda_n, u_n)$, $\|u_n\| = \varepsilon$ and $\lambda_n \to \lambda_0$. By the compactness of $G, \{u_n\}$ has a subsequence convergent to a solution of (9.5.1) with $\lambda = \lambda_0$ and which has norm ε in $\mathscr{B}$; this is a contradiction.

Theorem 9.5.4 *Suppose that $\lambda_0 \in r(L)$ is of odd multiplicity m and H is Fréchet-differentiable in u at $u = 0$. Then $(\lambda_0, 0)$ is a point of bifurcation.*

Proof If $(\lambda_0, 0)$ is not a point of bifurcation, then we can find an $\varepsilon > 0$ such that (9.5.1) with $\lambda = \lambda_0$ has no solutions of norm ε in $\mathscr{B}$. Choose δ as in Lemma 9.5.3. With $D = B(0, \varepsilon) \subset \mathscr{B}$, we consider the Leray–Schauder degree $d(I - G(\lambda, .), D, 0)$. Now by the conditions on $H, H_u(\lambda, u) = 0$ when $u = 0$. By Theorem 5.2.3, $i(I - \lambda L - H(\lambda, .), 0, 0) = i(I - \lambda L, 0, 0)$ if $\lambda \neq \lambda_0$; by Theorem 8.1.1, $i(I - \lambda L, 0, 0)$ changes by a multiplicative factor $(-1)^m$ as λ increases through the value λ_0. On the other hand, it follows from Lemma 9.5.3 and homotopy invariance that $d(I - G(\lambda, .), D, 0)$ is independent of λ in an interval $\lambda_0 - \delta < \lambda < \lambda_0 + \delta$. If $u = 0$ is the only solution of (9.5.1) with $\|u\| < \varepsilon, d(I - G(\lambda, .), D, 0)$ is just the index of $u = 0$. Since m is odd, it follows that there must be a non-trivial solution of (9.5.1) in $\{\|u\| < \varepsilon, |\lambda - \lambda_0| < \delta_1\}$ for all $\delta_1 < \delta$ – that is, $\lambda = \lambda_0$ is a bifurcation value after all.

When λ_0 is a simple characteristic value of L, it is known that the bifurcating solutions form, in the neighbourhood of $(\lambda_0, 0)$, a curve $(\lambda(s), u(s))$ in $\mathscr{E}$ passing through $(\lambda_0, 0)$ (see Sattinger (1973)). In this case, therefore, there are three possibilities for the bifurcating solutions: (*a*) there are two for $\lambda < \lambda_0$ and none for $\lambda > \lambda_0$, (*b*) there are two for $\lambda > \lambda_0$ and none for $\lambda < \lambda_0$, (*c*) there is one for $\lambda < \lambda_0$ and one for $\lambda > \lambda_0$. The index of each of these solutions is $+1$ or -1, for solutions are simple whenever $\lambda \neq \lambda_0$ and $|\lambda - \lambda_0|$ is small enough. As in the proof of Theorem 9.5.4, the index of the zero solution either changes from $+1$ to -1 or from -1 to $+1$ as λ increases through λ_0; furthermore, the total index of solutions near $u = 0$ is the same for $\lambda < \lambda_0$ as for $\lambda > \lambda_0$. If the zero solution has index $+1$, say, for $\lambda < \lambda_0$, we see that the following are the possibilities for the bifurcating solutions:

(1) there are two of index -1 for $\lambda < \lambda_0$ and none for $\lambda > \lambda_0$,

(2) there are two of index $+1$ for $\lambda > \lambda_0$ and none for $\lambda < \lambda_0$,

(3) there is one of index -1 for $\lambda < \lambda_0$ and one of index $+1$ for $\lambda > \lambda_0$.

There is a close connection between the index of a solution of (9.5.1) and its stability properties. When G is differentiable, we define a solution (λ_0, u_0) of (9.5.1) to be *linearly stable* if the eigenvalues of the operator $I - G_u(\lambda_0, u_0)$ all have positive real parts (G_u is the partial derivative of G with respect to u). Similarly (λ_0, u_0) is *linearly unstable* if at least one of the eigenvalues has negative real part.

Sattinger (1971) showed that if λ_0 is a simple bifurcation value for (9.5.1), and the zero solution is linearly stable for $\lambda < \lambda_0$ and linearly unstable for $\lambda > \lambda_0$, then the bifurcating solutions are linearly stable if they occur with $\lambda > \lambda_0$ and linearly unstable if they occur with $\lambda < \lambda_0$. This is accomplished by using the characterisation (1), (2) and (3) of the last paragraph but one; essentially, the solutions of index $+1$ are stable, while those of index -1 are unstable. This phenomenon is sometimes called 'exchange of stability'.

When bifurcation takes place at a multiple characteristic value of L the situation is less clear. There may be infinitely many bifurcating solutions or, when the multiplicity is even, there may be none. Bifurcation at a characteristic value of multiplicity 2 has been investigated by McLeod and Sattinger (1973); they show, in particular, that the above-mentioned result on the loss of stability at a characteristic value no longer holds.

We have required that G is a compact mapping for convenience, so that no more general form of degree than that of Leray–Schauder is needed. By using methods which are not degree theoretic, Crandall and Rabinowitz (1971) proved a general result on the local structure of the solutions of an equation such as (9.5.1) near a bifurcation point; we quote their main theorem for the sake of completeness.

Theorem 9.5.5 *Let $\mathscr{B}_1, \mathscr{B}_2$ be Banach spaces, Ω be an open subset of $\mathscr{B}_1$ and $G: \Omega \to \mathscr{B}_2$ be a twice continuously differentiable mapping. Suppose that $w: [-1, 1] \to \Omega$ is a simple, continuously differentiable curve in Ω such that $G(w(t)) = 0$ for $|t| \leqslant 1$. Suppose also that* (*a*) $w'(0) \neq 0$, (*b*) $\dim \ker \{G'(w(0))\} = 2$, $\operatorname{codim} \{\operatorname{Im} G'(w(0))\} = 1$, (*c*) *the space* $\ker \{G'(w(0))\}$ *is spanned by* $w'(0)$ *and* v, *say, and* (*d*)

$G''(w(0))\,(w'(0),v) \notin \operatorname{Im}\{G'(w(0))\}$. *Then* $w(0)$ *is a bifurcation point of* G *with respect to* $\mathscr{C} = \{w(t); t \in [-1, 1]\}$, *and in a neighbourhood of* $w(0)$ *the solutions of* $G(w) = 0$ *form two continuous curves intersecting only at* $w(0)$.

So far, the results have been of a local nature. We go on to prove some theorems on the solutions of (9.5.1) away from their bifurcation from the curve of trivial solutions (see Rabinowitz (1971)). Recall that a *continuum* is a closed, connected set. The notation is as above.

Lemma 9.5.6 *Let* $\mu \in r(L)$. *Suppose that every subcontinuum of* $\mathscr{S}_\mu = \mathscr{S} \cup \{(\mu, 0)\}$ *which meets* $(\mu, 0)$ *is bounded and contains no point* $(\lambda, 0)$ *with* $\mu \neq \lambda \in r(L)$. *Then there is a bounded, open subset* D *of* $\mathscr{E}$ *and* $\delta > 0$ *such that* (i) $(\mu, 0) \in D$, (ii) $\partial D \cap \mathscr{S} = \emptyset$ *and* (iii) D *contains no trivial solutions of* (9.5.1) *except those in* $B((\mu, 0), \delta)$, *where* $\delta < \varepsilon_0$, *the distance from* μ *to* $r(L)\backslash\{\mu\}$.

Proof Let $\mathscr{C}_\mu$ be the component of $(\mu, 0)$ in $\mathscr{S}_\mu$. By hypothesis, $\mathscr{C}_\mu$ is bounded, whence, by the compactness of G, it is also compact. Let N_δ be the δ-neighbourhood of $\mathscr{C}_\mu$; for δ sufficiently small, N_δ contains no solution $(\lambda, 0)$ with $|\lambda - \mu| > \delta$. Write $K = \bar{N}_\delta \cap \mathscr{S}$; K is a compact subset of $\mathscr{E}$. Since $\mathscr{C}_\mu \cap \partial N_\delta = \emptyset$, there are disjoint compact subsets K_1, K_2 of K such that $\mathscr{C}_\mu \subset K_1, \partial N_\delta \cap \mathscr{S} \subset K_2$, and $K = K_1 \cup K_2$. We can now choose D to be a neighbourhood of K_1.

Theorem 9.5.7 *Suppose that* $\mu \in r(L)$ *is of odd multiplicity. Then* $\mathscr{S}_\mu$ *contains a maximal subcontinuum* $\mathscr{C}_\mu$ *which contains* $(\mu, 0)$ *and which either* (i) *is unbounded, or* (ii) *contains a point* $(\lambda, 0)$ *with* $\mu \neq \lambda \in r(L)$.

Proof (i) It is convenient to write $\phi(\lambda, u) = u - G(\lambda, u)$, and ϕ_λ for the mapping $u \mapsto \phi(\lambda, u)$; thus the solutions of (9.5.1) are the zeros of ϕ. We suppose that no such $\mathscr{C}_\mu$ as is described in the enunciation of the theorem exists, and derive a contradiction. If there is no such $\mathscr{C}_\mu$, then there exist D, δ as in Lemma 9.5.6. Let $D_\lambda = \{u; (\lambda, u) \in D\}$.

(ii) For a fixed λ in the range $0 < |\lambda - \mu| \leqslant \delta, u = 0$ is an isolated solution of (9.5.1). Then there is $\eta(\lambda) > 0$ such that $(\lambda, 0)$ is the only solution in $\{\lambda\} \times \bar{B}(0, \eta(\lambda))$. If we define $\eta(\lambda) = \eta(\mu+\delta)$ for $\lambda > \mu + \delta$ and $\eta(\lambda) = \eta(\mu - \delta)$ for $\lambda < \mu - \delta$, we can arrange that $\bar{D}_\lambda \cap B(0, \eta(\lambda)) = \emptyset$ if $|\lambda - \mu| \geqslant \delta$ by choosing $\eta(\mu \pm \delta)$ sufficiently small. Thus, for $\lambda \neq \mu$, the Leray–Schauder degree $d(\phi_\lambda, V_\lambda, 0)$ is defined, where $V_\lambda = D_\lambda \backslash \bar{B}(0, \eta(\lambda))$.

(iii) Suppose that $\lambda > \mu$; so choose $\lambda_1 > \lambda$ that $(v, u) \in D$ implies that $v < \lambda_1$ (recall that D in bounded). Let $\eta = \inf\{\eta(v); \lambda \leqslant v \leqslant \lambda_1\}$, $B_\eta = B(0, \eta)$, and $D'_v = D_v \backslash \bar{B}_\eta$; clearly $\eta > 0$. Now, by construction, $0 \notin \phi_v(\partial D'_v)$; by the form of homotopy invariance of Theorem 4.3.15, $d(\phi_v, D'_v, 0)$ is independent of v in the interval $[\lambda, \lambda_1]$. But $D_{\lambda_1} = \emptyset$, whence $d(\phi_v, D'_v, 0) = 0$ for $v \in [\lambda, \lambda_1]$. Since $\phi_\lambda \neq 0$ in $(B(0, \eta(\lambda))\backslash \bar{B}_\eta)$, the excision property of degree leads to

$$d(\phi_\lambda, V_\lambda, 0) = 0 \quad (\lambda > \mu). \tag{9.5.3}$$

A similar argument works if $\lambda < \mu$.

(iv) By homotopy invariance, we also know that $d(\phi_\lambda, D_\lambda, 0)$ is independent of λ in an interval $|\lambda - \mu| < \varepsilon$. Take λ_2 and λ_3 to satisfy $\mu - \varepsilon < \lambda_2 < \mu < \lambda_3 < \mu + \varepsilon$. Since $u = 0$ is an isolated zero of ϕ_λ if $\lambda \notin r(L)$, we can define the index of the zero solution – $i(\lambda)$, say – for $0 < |\lambda - \mu| < \varepsilon$. By the domain decomposition property of degree,

$$\left.\begin{aligned} &d(\phi_{\lambda_2}, D_{\lambda_2}, 0) = d(\phi_{\lambda_2}, V_{\lambda_2}, 0) + i(\lambda_2) \\ \text{and}\quad &d(\phi_{\lambda_3}, D_{\lambda_3}, 0) = d(\phi_{\lambda_3}, V_{\lambda_3}, 0) + i(\lambda_3). \end{aligned}\right\} \tag{9.5.4}$$

(v) Equations (9.5.3) and (9.5.4) together imply that

$$d(\phi_v, D_v, 0) = i(v) \quad (v = \lambda_2, \lambda_3).$$

Thus $i(v)$ is independent of v in $0 < |v - \mu| < \varepsilon$. But μ is a characteristic value of L of odd multiplicity, so that $i(\lambda_2) \cdot i(\lambda_3) = -1$. This contradiction completes the proof.

Remark Both possibilities of Theorem 9.5.7 can occur. The unbounded possibility does not, of course, happen in the many cases where an *a priori* bound for solutions of (9.5.1) can be obtained. If it is known that (i) of the theorem is ruled out for a particular equation, we can obtain more detailed information about the set of bifurcation values.

Theorem 9.5.8 *Suppose that the hypotheses of Theorem* 9.5.7 *are satisfied, but that* (i) *of that theorem does not occur. Let* $\Gamma = \{v \in r(L);\ (v,0) \in \mathscr{C}_\mu, v \neq \mu\}$. *Then* Γ *contains at least one characteristic value of odd multiplicity.*

Proof Γ consists of a finite number of points: $v_1 < v_2 < \ldots < v_k$, say. By an argument similar to that of Lemma 9.5.6, $\delta > 0$ and a subset D of $\mathscr{E}$ can be found such that (i) $D \supset \mathscr{C}_\mu$, (i)) $\partial D \cap \mathscr{S} = \emptyset$, (iii) D contains no trivial solutions other than those within δ of μ or some member of Γ. Here $\delta < \varepsilon_1$, the distance from $\Gamma \cup \{\mu\}$ to the rest of $r(L)$. The notation throughout the proof is that used in Lemma 9.5.6 and Theorem 9.5.7. We define D_λ and $\eta(\lambda)$ as before: $\eta(\lambda)$ is constant outside the δ-neighbourhoods of μ and $v_i (i = 1, \ldots, k)$.

Suppose that none of the v_i is of odd multiplicity. Referring back to (9.5.4), since the left hand sides of the two equations are equal, and $i(\lambda_2)$ and $i(\lambda_3)$ are of opposite signs, we see that at least one of $d(\phi_{\lambda_2}, V_{\lambda_2}, 0), d(\phi_{\lambda_3}, V_{\lambda_3}, 0)$ is non-zero.

Let v_s be the smallest member of Γ which is greater than μ. Since v_s is of even multiplicity, the argument we used in the proof of Theorem 9.5.7 yields

$$d(\phi_{\lambda_4}, V_{\lambda_4}, 0) = d(\phi_{\lambda_5}, V_{\lambda_5}, 0)$$

if $v_s - \varepsilon < \lambda_4 < v_s < \lambda_5 < v_s + \varepsilon$. Also $d(\phi_\lambda, V_\lambda, 0)$ is independent of λ in the interval (v_s, v_{s+1}). Continuing in this way, and observing that $D_\lambda = \emptyset$ for λ sufficiently large, we find that

$$d(\phi_\lambda, V_\lambda, 0) = 0 \quad (\lambda > \mu, \lambda \notin \Gamma).$$

A similar argument shows that

$$d(\phi_\lambda, V, 0) = 0 \quad (\lambda < \mu, \lambda \notin \Gamma).$$

But we have already noted that $d(\phi_\lambda, V_\lambda, 0)$ is non-zero for at least one of $\lambda = \lambda_2$ and $\lambda = \lambda_3$. This contradiction establishes the result.

When μ is a characteristic value of L of multiplicity 1, a good deal more can be said about the structure of $\mathscr{C}_\mu$. If it is supposed that G is differentiable, then it can be shown that $\mathscr{C}_\mu$ is the union of two subcontinua containing $(\mu, 0)$ and having no other common point in a neighbourhood of $(\mu, 0)$. Degree theoretic arguments are used by Rabinowitz (1971) to show that each of the subcontinua is either unbounded or meets the curve of trivial solutions in a point $(\hat{\mu}, 0)$, with $\mu \neq \hat{\mu} \in r(L)$.

The foregoing strongly suggests that similar results exist for mappings G which are non-compact but for which a degree can be defined. This is indeed so. Stuart (1973) develops a bifurcation theory along the lines of the above for equations $u = \lambda G(u) = \lambda(L(u) + H(u))$ when G is a strict set contraction; this uses the degree developed in Chapter 6. Still more generally, Dancer (1975) considered equations $u = G(\lambda, u)$, where G is differentiable and, in the notation of Section 8.2, $I - G \in \mathscr{F}(\mathscr{B})$; the degree defined in that section is used. We shall not enter into the details of these extensions, but refer the reader to the original papers.

'Bifurcation theory' as outlined above is an abstract theory which has many applications – as is seen in the references which we have quoted. These range from non-linear Sturm–Liouville type problems in ordinary differential equations to eigenvalue problems for elliptic partial differential equations. Such problems are reduced to operator equations of the form (9.5.1), usually in a very natural way. The interested reader will rapidly locate a large collection of these applications in the literature.

9.6 Other applications

We conclude with a short section in which we give an indication of some ways in which degree theory can be used in areas we have not, so far in this book, encountered. Although little specific mention has been made of partial differential equations, this does not mean that topological degree is an unimportant tool in their study. The point is that partial differential equations are often converted into operator equations in a suitable infinite dimensional normed space; general techniques developed in this setting may then be used. Some specific examples can be found in Cronin (1964, 1974), Berger and Berger (1968) and Rabinowitz (1971); these sources also contain further references. Similarly, many problems involving functional differential equations (including differential equations with delays) are treated by converting them into operator equations.

Many of the ideas of earlier parts of this chapter have fairly obvious analogues which are applicable in this more general context.

We shall briefly sketch two examples of functional differential equations. (For background, see Hale (1971).) The degree of multi-valued mappings defined in Section 7.3 can be used to investigate the following boundary value problem (Schmitt, 1970):

$$\left.\begin{aligned} Lx(t) &\equiv x''(t) + a(t)x'(t) + b(t)x(t) + c(t)x(t-d(t)) \\ &= e(t) \quad (0 \leqslant t \leqslant T), \\ x(t) &= \phi(t) \quad (t \notin [0, T]). \end{aligned}\right\} \quad (9.6.1)$$

Here $a, b, c, d, e, \phi \in \mathscr{P}_T$, the space of real, continuous functions of period T, with supremum norm. For $\alpha, \beta \in \mathscr{P}_T$, let

$$[\alpha, \beta] = \{\psi \in \mathscr{P}_T; \alpha(t) \leqslant \psi(t) \leqslant \beta(t), t \in [0, T]\}.$$

If $L\beta(t) \leqslant e(t) \leqslant L\alpha(t)$ for $t \in [0, T]$, we denote by $S(\phi)$ the set of solutions of (9.6.1). It can be shown that for $\phi \in [\alpha, \beta], S(\phi)$ is a non-empty, convex subset of $[\alpha, \beta]$; the multivalued mapping S is upper semicontinuous. A compact, convex subset M of $\mathscr{P}_T$ can be found so that $S(\phi) \subset M$ if $\phi \in M$. The mapping S therefore has a fixed point – that is, there is ψ such that $\psi \in S(\psi)$; such a fixed point corresponds to a periodic solution of (9.6.1). The same conclusion holds whenever a suitable set $D \subset \mathscr{P}_T$ can be found with the property that $d(I - S, D, 0) \neq 0$ (Theorem 7.3.6).

Our second example is a neutral functional differential equation (Hale and Mawhin, 1974). For $r > 0$, let C be the space of continuous functions from $[-r, 0]$ into $\mathbf{R}^n$ with the supremum norm – in the obvious notation, $C = C([-r, 0], \mathbf{R}^n)$. If $\delta > 0$ and $x \in C([\sigma - r, \sigma + \delta], \mathbf{R}^n)$, define x_t to be the translate of x through t:

$$x_t(\theta) = x(t + \theta) \quad (\theta \in (-r, 0]).$$

With $\omega > 0, \mathscr{P}_\omega$ is the space of functions $\mathbf{R} \to \mathbf{R}^n$ of period ω, with the supremum norm. Define $D: \mathbf{R} \times C \to \mathbf{R}^n$ by

$$D(t)\phi = \phi(0) - A(t, \phi),$$

where $A: \mathbf{R} \times C \to \mathbf{R}^n$ is continuous, periodic in t, linear in ϕ and satisfies an inequality $|A(t, \phi)| \leqslant \gamma(s) \| \phi \|$ if $\phi \in C$ and $\phi = 0$ on

$[-r, -s]$. Further, define the mapping $L: \mathscr{P}_\omega \to C$ by

$$Lx(t) = D(t)x_t - D(0)x_0.$$

D is said to be *stable* if the zero solution of the equation $D(t)y_t = 0$ is uniformly asymptotically stable. Hale and Mawhin show that, if D is stable, then L is a Fredholm map of index zero.

Periodic solutions are sought of the equation

$$\frac{d}{dt} D(t)x_t = \frac{d}{dt} x(t) + f(t, x_t),$$

where f is continuous, periodic in t, and maps bounded sets to bounded sets. It can be easily shown that the mapping $N: C \to C$ given by

$$Nx(t) = \int_0^t f(s, x_s)\, ds - x(0)$$

is compact. The problem has therefore been reduced to solving $Lx = x + Nx$ in the space $\mathscr{P}_\omega$. We can apply the degree theory developed in Chapter 8 to this problem: there is a solution if a bounded, open set D can be found so that $d(I + N - L, D, 0) \neq 0$. The calculation of this degree can be difficult; it should be reduced to a calculation in a finite dimensional space, as in Sections 8.1 and 5.2.

One other application of degree theory, which we have not touched upon in this chapter, is to do with 'mapping theorems'. These are results similar to the theorem of domain invariance of Chapter 3; such results are more difficult to prove, of course, in infinite dimensional spaces. Particular cases can be found, for example, in Nussbaum (1972) and Petryshyn (1970). Strict set contractions are considered in the former, and A-proper mappings in the latter.

References

Altman, M. (1957): A fixed point theorem in Banach space. *Bull. Acad. Polon. Sci. Cl. III* **5**, 89–92.

Amann, H. and Weiss, S. A. (1973): On the uniqueness of the topological degree. *Math. Z.* **130**, 39–54.

Bebernes, J. W. (1974): A simple alternative problem for finding periodic solutions of second order ordinary differential systems. *Proc. Amer. Math. Soc.* **42**, 121–7.

Bebernes, J. W. and Schmitt, K. (1973): Periodic boundary value problems for systems of second order differential equations. *J. Differential Equations* **13**, 32–47.

Berger, M. S. and Berger, M. S. (1968): *Perspectives in nonlinearity*. Benjamin, New York.

Borsuk, K. (1933): Drei Sätze über die n-dimensionale euklidische Sphäre. *Fund. Math.* **20**, 177–90.

Brouwer, L. E. J. (1912): Über Abbildung von Mannigfaltigkeiten, *Math. Ann.* **71**, 97–115.

Browder, F. E. (1968): Topology and non-linear functional equations. *Studia Math.* **31**, 189–204.

Browder, F. E. and Gupta, C. P. (1969): Topological degree and non-linear mappings of analytic type in Banach spaces. *J. Math. Anal. Appl.* **26**, 390–402.

Browder, F. E. and Nussbaum, R. D. (1968): The topological degree for noncompact nonlinear mappings in Banach spaces. *Bull. Amer. Math. Soc.* **74**, 671–6.

Browder, F. E. and Petryshyn, W. V. (1968): Noncompact operators in Banach spaces. *Bull. Amer. Math. Soc.* **74**, 641–6.

Browder, F. E. and Petryshyn, W. V. (1969): Approximation methods and the generalised topological degree for nonlinear mappings in Banach spaces. *J. Functional Analysis* **3**, 217–45.

Cellina, A. (1969): Approximation of set valued functions and fixed point theorems. *Ann. Mat. Pura Appl.* (4) **82**, 17–24.

Cellina, A. (1971): On the existence of solutions of ordinary differential equations in Banach spaces. *Funkcial. Ekvac.* **14**, 129–36.

Cellina, A. and Lasota, A. (1969): A new approach to the definition of topological degree for multi-valued mappings. *Atti Accad. Naz. Lincei Rend. Cl. Sci. Fis. Mat. Natur.* **47**, 434–40.

Coddington, E. A. and Levinson, N. (1955): *Theory of ordinary differential equations*. McGraw-Hill, New York.

Crandall, M. G. and Rabinowitz, P. H. (1971): Bifurcation from simple eigenvalues. *J. Functional Analysis* **8**, 321–40.

Crandall, M. G. and Rabinowitz, P. H. (1973): Bifurcation, perturbation of

simple eigenvalues, and linearised stability. *Arch. Rational Mech. Anal.* **52**, 161–80.
Cronin, J. (1964): *Fixed points and topological degree in nonlinear analysis.* Amer. Math. Soc., Providence, R. I.
Cronin, J. (1965): The point at infinity and periodic solutions. *J. Differential Equations* **1**, 156–70.
Cronin, J. (1967): Periodic solutions of some non-linear differential equations. *J. Differential Equations* **3**, 31–46.
Cronin, J. (1971): Topological degree and the number of solutions of equations. *Duke Math. J.* **38**, 531–8.
Cronin, J. (1974): Quasilinear equations and equations with large nonlinearities. *Rocky Mountain J. Math.* **4**, 41–63.
Dancer, E. N. (1975): Boundary-value problems for ordinary differential equations on infinite intervals. *Proc. London Math. Soc.* (3) **30**, 76–94.
Darbo, G. (1955): Punti uniti in trasformazioni a codiminio non compatto. *Rend. Sem. Mat. Univ. Padova* **24**, 84–92.
de Figueiredo, D. G. (1967): Fixed point theorems for nonlinear operators and Galerkin approximations. *J. Differential Equations* **3**, 271–81.
Deimling, K. (1970): Fixed points of generalised P-compact operators. *Math. Z.* **115**, 188–96.
Dieudonné, J. (1960): *Foundations of modern analysis.* Academic Press, London.
Dugundji, J. (1951): An extension of Tietze's theorem. *Pacific J. Math.* **1**, 353–67.
Dunford, N. and Schwartz, J. T. (1958): *Linear operators*, Part I. Interscience, New York.
Edmunds, D. E., Potter, A. J. B. and Stuart, C. A. (1972): Non-compact positive operators. *Proc. Roy. Soc. London Ser. A* **328**, 67–81.
Eggleston, H. G. (1958): *Convexity.* Cambridge University Press.
Eilenberg, S, and Montgomery, D. (1946): Fixed point theorems for multi-valued transformations. *Amer. J. Math.* **68**, 214–22.
Elworthy, K. D. and Tromba, A. J. (1970): Degree theory on Banach manifolds. In *Proceedings of symposia in pure mathematics*, Vol. 18, Part 1. Amer. Math. Soc., Providence, R. I., 86–94.
Ezeilo, J. O. C. (1975): A further result on the existence of periodic solutions of the equation $\dddot{x} + a\ddot{x} + b\dot{x} + h(x) = p(t,x,\dot{x},\ddot{x})$. *Math. Proc. Cambridge Philos. Soc.* **77**, 547–51.
Fenske, C. (1971): Analytische Theorie des Abbildungsgrades für Abbildungen in Banachräumen. *Math. Nachr.* **48**, 279–90.
Fitzpatrick, P. M. (1972): A-proper mappings and their uniform limits. *Bull. Amer. Math. Soc.* **78**, 806–9.
Fraenkel, L. E. (1973): Lecture notes, Cambridge University. Unpublished.
Furi, M. and Vignoli, A. (1970): On α-nonexpansive mappings and fixed points. *Atti Accad. Naz. Lincei Rend. Cl. Sci. Fis. Mat. Natur.* **48**, 195–8.
Gokhberg, I. C., Goldenstein, L. S. and Markus, A. S. (1957): Investigation of some properties of bounded linear operators in connection with their q-norms. *Uchen. Zap. Kishinev. Gos. Univ.* **29**, 29–36. (In Russian.)
Gokhberg, I. C. and Krein, M. G. (1960): The basic propositions on deficiency numbers, root numbers and indices of linear operators. *Amer. Math. Soc. Transl.* (2) **13**, 185–264.

Gomory, R. E. (1956): Critical points at infinity and forced oscillations. *Contributions to the theory of non-linear oscillations*, Vol. 3 (Ann. of Math. Studies 36), 85–126.
Granas, A. (1959*a*): Theorem on antipodes and theorems on fixed points for a certain class of multi-valued mappings in Banach spaces. *Bull. Acad. Pol. Sci. Sér Sci. Math. Astronom. Phys.* **7**, 271–5.
Granas, A. (1959*b*): Sur la notion du degré topologique pour une certaine classe de transformations multivalentes dans les espaces de Banach. *Bull. Acad. Polon. Sci. Sér. Sci. Math. Astronom. Phys.* **7**, 191–4.
Granas, A. and Jaworowski, J. W. (1959): Some theorems on multivalued mappings of subsets of Euclidean space, *Bull. Acad. Polon. Sci. Sér. Sci. Math. Astronom. Phys.* **7**, 277–83.
Gronwall, T. H. (1919): Note on the derivatives with respect to a parameter of the solutions of a system of differential equations. *Ann. Math.* (2) **20**, 292–6.
Grothendieck, A. (1973): *Topological vector spaces.* Gordon and Breach, New York.
Halanay, A. (1966): *Differential equations.* Academic Press, New York.
Hale, J. K. (1971): *Functional differential equations.* Springer, New York.
Hale, J. K. and Mawhin, J. (1974): Coincidence degree and periodic solutions of neutral equations. *J. Differential Equations* **15**, 295–307.
Heinz, E. (1959): An elementary analytic theory of the degree of mapping in n-dimensional space. *J. Math. and Mech.* **8**, 231–47.
Hervé, M. (1963): *Several complex variables.* TATA Institute, Bombay.
Hopf, H. (1926a): Abbildungsklassen n-dimensionaler Mannigfaltigkeiten. *Math. Ann.* **96**, 209–24.
Hopf, H. (1926*b*): Vektorfelder in n-dimensionalen Mannigfaltigkeiten. *Math. Ann.* **96**, 225–50.
Hukuhara, M. (1967): Sur l'application semi-continue dont la valeur est un compact convexe. *Funkcial. Ekvac.* **10**, 43–66.
Kelley, J. L. (1955): *General topology.* Van Nostrand, Princeton.
Knobloch, H. W. (1971): On the existence of periodic solutions of second order vector differential equations. *J. Differential Equations* **9**, 67–86.
Kotin, L. (1968): A Floquet theorem for real nonlinear systems. *J. Math. Anal. Appl.* **21**, 384–8.
Krasnosel'skii, M. A. (1955): Two remarks on the method of successive approximations. *Uspehi Mat. Nauk* **10**, 123–7. (In Russian.)
Krasnosel'skii, M. A. (1964): *Topological methods in the theory of nonlinear integral equations.* Pergamon, Oxford.
Kuratowski, K. (1930): Sur les espaces complets. *Fund. Math.* **15**, 301–9.
Lando, C. A. (1971): Periodic solutions of nonlinear systems with forcing terms. *J. Differential Equations* **9**, 262–79.
Lazer, A. C. (1968): On Schauder's fixed point theorem and forced second-order nonlinear oscillations. *J. Math. Anal. Appl.* **21**, 421–5.
Lefschetz, S. (1952): Notes on differential equations. *Contributions to the theory of nonlinear oscillations*, Vol. 2 (Ann. of Math. Studies 29) 61–74.
Leray, J. (1935): Topologie des espaces de Banach. *C. R. Acad. Sci. Paris* **200**, 1082–4.
Leray, J. (1936): Les problèmes non linéaires. *Enseignement Math.* **35**, 139–51.

Leray, J. (1950): La théorie des points fixes et ses applications en analyse. *Proceedings of the International Congress of Mathematicians*, Vol. 2, 202–8.
Leray, J. and Schauder, J. (1934): Topologie et équations fonctionnelles. *Ann. Sci. École Norm. Sup. Sér. 3* **51**, 45–78.
Lloyd, N. G. (1975): On analytic differential equations. *Proc. London Math. Soc.* (3) **30**, 430–44.
Ma, T. W. (1972): Topological degrees for set-valued compact vector fields in locally convex spaces. *Dissertationes Math. (Rozprawy Mat.)* **92**, 1–43.
Mawhin, J. (1970): Existence of periodic solutions for higher order differential systems that are not of class *D*. *J. Differential Equations* **8**, 523–30.
Mawhin, J. (1972): Equivalence theorems for nonlinear operator equations and coincidence degree theory for some mappings in locally convex topological vector spaces. *J. Differential Equations* **12**, 610–36.
McLeod, J. B. and Sattinger, D. H. (1973): Loss of stability and bifurcation at a double eigenvalue, *J. Functional Analysis* **14**, 62–84.
Milnor, J. (1965): *Topology from a differentiable viewpoint.* University Press of Virginia, Charlottesville.
Nagumo, M. (1951*a*): A theory of degree of mapping based on infinitesimal analysis. *Amer. J. Math.* **73**, 485–96.
Nagumo, M. (1951*b*): Degree of mapping in convex linear topological spaces. *Amer. J. Math.* **73**, 497–511.
Nemytskii V. V. and Stepanov, V. V. (1960): *Qualitative theory of differential equations.* Princeton University Press.
Newman, M. H. A. (1939): *Elements of the topology of plane sets of points.* Cambridge University Press.
Nussbaum, R. D. (1969): The fixed point index and asymptotic fixed point theorems for *k*-set contractions. *Bull. Amer. Math. Soc.* **75**, 490–5.
Nussbaum, R. D. (1971*a*): The fixed point index for local condensing maps. *Ann. Mat. Pura Appl.* (4) **89**, 217–58.
Nussbaum, R. D. (1971*b*): Estimates for the number of solutions of operator equations. *Applicable Anal.* **1**, 183–200.
Nussbaum, R. D. (1972): Degree theory for local condensing maps. *J. Math. Anal. Appl.* **37**, 741–66.
Petryshyn, W. V. (1966): On nonlinear *P*-compact operators in Banach space with applications to constructive fixed point theorems. *J. Math. Anal. Appl.* **15**, 228–42.
Petryshyn, W. V. (1970): Invariance of domain theorem for locally *A*-proper mappings and its implications. *J. Functional Analysis* **5**, 137–59.
Petryshyn, W. V. (1971): Structure of the fixed point sets of *k*-set contractions. *Arch. Rational Mech. Anal.* **40**, 312–28.
Petryshyn, W. V. (1973): Fixed point theorems for various classes of 1-set contractive and 1-ball contractive mappings in Banach spaces. *Trans. Amer. Math. Soc.* **182**, 323–52.
Petryshyn, W. V. and Fitzpatrick, P. M. (1974): A degree theory, fixed point theorem, and mapping theorems for multivalued noncompact mappings. *Trans. Amer. Math. Soc.* **194**, 1–25.
Petryshyn, W. V. and Fitzpatrick, P. M. (1975): Fixed point theorems and the fixed point index for multivalued mappings in cones. *J. Lond. Math. Soc.* (2) **12**, 75–85.

Petryshyn, W. V. and Tucker, T. S. (1969): On the functional equations involving nonlinear generalised P-compact operators. *Trans. Amer. Math. Soc.* **135**, 343–73.
Pontryagin, L. S. (1959): Smooth manifolds and their applications in homotopy theory. *Amer. Math. Soc. Translations, Ser. 2* **11**, 1–114. (Translated from *Trudy Inst. Steklov* **45** (1955).)
Rabinowitz, P. H. (1971): Some global results for nonlinear eigenvalue problems. *J. Functional Analysis* **7**, 487–513.
Reich, S. (1972): Fixed points in locally convex spaces. *Math. Z.* **125**, 17–31.
Reissig, R. (1970): Periodic solutions of a nonlinear nth-order vector differential equation. *Ann. Mat. Pura Appl.* (4) **87**, 111–24.
Reissig, R. (1972*a*): Periodic solutions of a third order nonlinear differential equation. *Ann. Mat. Pura Appl.* (4) **92**, 193–8.
Reissig, R. (1972*b*): An extension of Ezeilo's result. *Ann. Mat. Pura Appl.* (4) **92**, 199–210.
Remmert, R. and Stein, K. (1953): Über die wesentlichen Singularitäten analytischer Mengen, *Math. Ann.* **126**, 263–306.
Rickart, C. E. (1960): *General theory of Banach algebras.* Van Nostrand, Princeton, N.J.
Robertson, A. P. and Robertson, W. J. (1966): *Topological vector spaces.* Cambridge University Press.
Rothe, E. (1937): Zur Theorie der topologischen Ordnung und der Vektorfelder in Banachschen Räumen. *Compositio Math.* **5**, 177–97.
Rudin, W. (1973): *Functional analysis.* McGraw-Hill, New York.
Sadovskii, B. N. (1967): On a fixed point principle. *Functional Anal. Appl.* **1**, 74–76.
Sard, A. (1942): The measure of the critical values of differentiable maps. *Bull. Amer. Math. Soc.* **48**, 883–90.
Sattinger, D. H. (1971): Stability of bifurcating solutions by Leray–Schauder degree. *Arch. Rational Math. Anal.* **43**, 154–66.
Sattinger, D. H. (1973): *Topics in stability and bifurcation theory* (Lecture Notes in mathematics, No. 309). Springer-Verlag, Berlin.
Schaefer, H. (1955): Über die Methode der a priori Schranken. *Math. Ann.* **129**, 415–16.
Schauder, J. (1930): Der Fixpunktsatz in Funktionalräumen. *Studia Math.* **2**, 171–80.
Schmitt, K. (1970): Periodic solutions of linear second order differential equations with deviating argument. *Proc. Amer. Math. Soc.* **26**, 282–5.
Schmitt, K. (1972): Periodic solutions of small period of systems of nth order differential equations. *Proc. Amer. Math. Soc.* **36**, 459–63.
Schwartz, J. T. (1963): Compact analytic mappings of B-spaces and a theorem of Jane Cronin. *Comm. Pure Appl. Math.* **16**, 253–60.
Schwartz, J. T. (1969): *Nonlinear functional analysis.* Gordon and Breach, New York.
Smale, S. (1965): An infinite dimensional generalisation of Sard's theorem. *Amer. J. Math.* **87**, 861–6.
Smart, D. R. (1974): *Fixed point theorems.* Cambridge University Press.
Sternberg, S. (1964): *Lectures on differential geometry.* Prentice-Hall, New York.
Stuart, C. A. (1973): Some bifurcation theory for k-set contractions. *Proc. London Math. Soc.* (3) **27**, 531–50

Stuart, C. A. and Toland, J. F. (1973): The fixed point index of a linear k-set contraction. *J. London Math. Soc.* (2) **6**, 317–20

Thomas, J. W. (1973): A bifurcation theorem for k-set contractions. *Pacific J. Math.* **44**, 749–56.

Thomas, J. W. (1974): Upper and lower bounds for the number of solutions of functional equations involving k-set contractions. *Rocky Mountain J. Math.* **4**, 89–93.

Tietze, H. (1915): Über Funktionen die auf einer abgeschlossenen Menge stetig sind. *J. Reine Angew. Math.* **145**, 9–14.

Tychonoff, A. (1935): Ein Fixpunktsatz. *Math. Ann.* **111**, 767–76.

Webb, J. R. L. (1971): Remarks on k-set contractions. *Boll. Un. Mat. Ital.* (4) **4**, 614–29.

Webb, J. R. L. (1974): On degree theory for multivalued mappings and applications, *Boll. Un. Mat. Ital.* (4) **9**, 137–58.

Webb, J. R. L. (1975); On uniqueness of topological degree for set-valued mappings. *Proc. Roy. Soc. Edinburgh* **74**A, 225–9.

Willard, S. (1970): *General topology.* Addison-Wesley, Reading, Massachusetts.

Wong, H. S. F. (1971): The topological degree of A-proper maps. *Canad. J. Math.* **23**, 403–12.

Index